D0759959

Reputable Conduct
Ethical Issues in Policing and Corrections

Second Edition

John R. Jones, M.Ed., Ph.D.

Law and Justice Centre,
Sir Sandford Fleming College

Daniel P. Carlson, B.S.

Center for Law Enforcement Ethics
Southwestern Law Enforcement Institute

Upper Saddle River, New Jersey 07458

Library of Congress Cataloging-in-Publication Data

Jones, John R., M. Ed., Ph. D.
Reputable conduct : ethical issues in policing and corrections / John R. Jones, Daniel P. Carlson.
p. cm.
Includes index.
ISBN 0-13-028620-6
1. Police ethics. 2. Corrections—Moral and ethical aspects. 3. Criminal justice, administration of—Moral and ethical aspects. I. Carlson, Daniel P. II. Title.

HV7924 .J66 2001
174'.93632—dc21 00-034651

Publisher: Dave Garza
Senior Acquisitions Editor: Kim Davies
Production Editor: Lori Dalberg, Carlisle Publishers Services
Production Liaison: Barbara Marttine Cappuccio
Director of Manufacturing and Production: Bruce Johnson
Managing Editor: Mary Carnis
Manufacturing Buyer: Ed O'Dougherty
Art Director: Marianne Frasco
Cover Designer: Miguel Ortiz
Marketing Manager: Chris Ruel
Editorial Assistant: Lisa Schwartz
Interior Design: Monica Kompter
Composition: Carlisle Communications, Ltd.
Printing and Binding: R. R. Donnelley & Sons

Prentice-Hall International (UK) Limited, *London*
Prentice-Hall of Australia Pty. Limited, *Sydney*
Prentice-Hall Canada Inc., *Toronto*
Prentice-Hall Hispanoamericana, S.A., *Mexico*
Prentice-Hall of India Private Limited, *New Delhi*
Prentice-Hall of Japan, Inc., *Tokyo*
Prentice-Hall Singapore Pte. Ltd.
Editora Prentice-Hall do Brasil, Ltda., *Rio de Janeiro*

10 9 8 7 6 5 4 3 2 1
ISBN 0-13-028620-6

To my wife Marilyn, and my three children, Steffan, Christopher, and Melissa, who have brought more love and meaning to my life than words can express. This book is also dedicated to the memory of my mother, Elizabeth Jones, whose generosity of spirit has left an indelible mark on my life.

John Jones

To Bonnie, who has been my high school sweetheart, my friend, my comforter, and my wife. She, our three children, and our grandchildren are the center of my world. This is dedicated, with love and appreciation, to them.

Dan Carlson

Table of Contents

Preface

Written for students in college criminal justice programs, and for police and corrections officers in training, *Reputable Conduct* has been designed as a working text rather than a reference book. And while this work combines Dr. Jones' 17 years of experience teaching ethics with Mr. Carlson's 33 years as a law enforcement practitioner and instructor, both acknowledge having been well taught by their many students and colleagues during that time. By design, the writing style and tone of this book are very personal, for as we wrote it, we imagined our students listening in.

Reputable Conduct is intended to be a friendly and easy-to-read introduction to the difficult and sometimes peculiar ethical demands of the professions of policing and corrections. It addresses some of the characteristics of these roles, with particular emphasis on subcultural constraints, and how loyalty to our colleague group can sometimes cause us to sacrifice our individuality. How these constraints may affect the moral decision making of an officer is looked at in detail in Chapters 5, 6, and 7.

One unusual feature of the book is the discussion in Chapter 4, which encourages the student to think about the role of the ethics educator. The discussion has been included so that the educator's role can be more clearly understood by the student. I hope this will reduce the possibility of misunderstanding arising between teacher and student regarding the purpose of an ethics course.

Tools that may be helpful for resolving moral dilemmas are discussed in Chapter 8. Chapter 9 includes several case studies to give students an opportunity to practice using these tools.

Some of the material in the book could be described as sensitive—even controversial—in nature. To gain maximum benefit from the experience of working through the book, students are encouraged to consider with an open mind the readings and the Reflections, which are searching questions interspersed throughout each chapter and based on the preceding content. This will require a degree of maturity and discipline.

Reputable Conduct is designed to provide students with a vehicle to promote private thought and class discussion about issues that—from what I have been told by my students and many practitioners over the years—are important to the vital roles played in our society by police and corrections officers.

If such thinking and discussion contribute, even in some small way, to justice continuing to be served, and to the individual officer's sense of well-being being enhanced, then the effort on all our parts will have been justified.

Acknowledgments

It has been a pleasure to work with staff members of Prentice Hall Canada Career & Technology in the writing of this book, my first attempt at such an enterprise. They have been unfailingly helpful, cooperative, and encouraging.

I wish to acknowledge the early and continuing work of David Stover, acquisitions editor. David has been a gracious contract negotiator and guide. His early suggestions for improving the book were invaluable. Ivetka Vasil, editorial assistant, undertook the review process and kept me informed in a most supportive way. Andrew Winton, production editor, shepherded the book through the production phase and did so with class and a little firmness when it was required. And I wish to acknowledge the expert help and advice of Allyson Latta, whose copyediting contributed greatly to improving the original manuscript.

Finally, I wish to say how much I valued the feedback received from the external reviewers: Janet Hoffman, Lambton College; Lori Larsen, Lethbridge Community College; Ann Parks, Lethbridge Community College; Rebecca Volk, Algonquin College; Megan Way Nicholson, St. Lawrence College; and Lisa Bezaire, Sir Sandford Fleming College. Their encouraging and constructive comments gave me the will to finish the book, and much food for thought. Some of their suggestions have been incorporated in this edition, and others have been kept on file for inclusion in what, I hope, will be subsequent editions of this book.

J.R.J.

About the Authors

John Jones

Dr. John Jones has worked for more than 30 years in the human development field, as minister, hospital chaplain, youth leader, correctional officer, high school teacher, probation officer and correctional institution administrator. During the last 17 years he has worked as an administrator and professor in the Law and Justice Centre at a Canadian college of applied arts and technology, where he has enjoyed teaching ethics to justice students.

John has also been active in training and consulting in the justice field. He has trained several thousand participants in Ethics in the Workplace, Team Development with an Ethical Twist, and Hostage Survival Skills. He is considered to be a caring, compassionate, humorous and skilled facilitator, seminar leader and speaker.

John holds a joint Honours Bachelor of Arts degree in psychology and philosophy from the University of Wales, Cardiff; a Master of Education degree from the University of Toronto (The Ontario Institute for Studies in Education); and a Doctor of Philosophy degree from the University of Toronto.

John has lived and worked in Britain, Jamaica, Australia and Canada. He is married with three children.

Dan Carlson

During the course of a law enforcement career that began in 1967, Dan Carlson worked for several years as a police patrolman and then a deputy sheriff before joining the New York State Police. After 21 years of full-time law enforcement service, he retired from the state police at the rank of Captain—Assistant Director of Training, and moved fully into the field of training and education. Dan managed a Regional Police Academy in the state of Texas, formed a private management consulting and training organization, and in 1992 was selected to guide the formation and development of the Center for Law Enforcement Ethics, the first organization of its kind in the United States.

Dan has presented ethics training programs for criminal justice organizations across the United States and Canada, and has been a regular guest speaker before law enforcement groups across North America. Widely published in the areas of ethics and management, Dan has authored more than two dozen articles for a range of professional journals. He graduated from the State University of New York, and was the 1985 recipient of the George Searle Award for Excellence in Law Enforcement Training. Dan has served on the ethics committees of both the American Society of Law Enforcement Trainers and the International Association of Chiefs of Police.

Dan and his wife, Bonnie, reside in Texas, only a short drive from their three children and three grandchildren.

Reputable Conduct

Introduction

Perhaps the most important thing I learned from this class is that from now on I should always do the following:

- Keep an open mind about every issue.
- Try to look at an issue from as many perspectives as possible, before I form an opinion.
- Test my bias on issues about which I've formed an opinion.
- Realize that because everyone doesn't look at life's issues in the same way that I do, doesn't make them wrong and me right.
- I should not be ashamed to change my opinion on an issue when I had earlier felt the opposite.
- This class helped to make me aware of how gray many of life's issues are—there are few issues that are really black or white.

—Journal entry, justice student in ethics course

I think that an ethical person is someone who is consistent in doing the "right" thing. Being ethical consists of determining what is morally correct, and standing up for what you believe is right. I think an ethical person is a confident person. Someone who is not influenced by others from what is the morally correct action to be taken in a given situation. I see myself as an ethical person but lack the confidence to stand up for every moral issue or ethical dilemma that I encounter.

—Journal entry, justice student in ethics course

Chapter Objectives

After reading this chapter, you should be able to:

- *Explain the purpose of this book.*
- *Explain how we may have arrived at our values.*
- *Explain how we in the justice field may inherit values.*
- *Explain the notion of* ***subculture*** *in policing and corrections.*
- *Apply knowledge of subculture in the military to policing and corrections.*
- *Explain the notion of* ***de-individuation.***

Why This Book?

For the last 20 years I have taught a course in ethics to students in the Law and Justice Centre of a Canadian college of applied arts and technology. For an equal number of years I have inquired of visiting publishers' representatives regarding texts available with respect to ethics education in the criminal justice field. The answer has always been that they know of little in this area. In addition, some of my faculty colleagues in other colleges, and some police and corrections training personnel, have asked me about such texts. On several occasions I was asked a similar question during a 1996 lecture tour of Australia.

That these inquiries are being made is encouraging, for there appears to be a growing interest in ethics training in the justice field. But the inquiries, along with some direct personal requests of me to do something concrete to meet the need, have translated into a personal challenge. You are now reading my response to that challenge.

What Kind of Book Is This?

This book is not intended to be a theoretical discourse on ethics. Background historical and theoretical information on ethics can be found, for example, in virtually any medical or business ethics text. This information is fairly consistently addressed in these texts and I think it would be redundant to cover similar in-depth material here. Chapter 2 does take a cursory look at the theory and history of moral philosophy, but readers wanting a more in-depth treatment may wish to consult other sources. This book is designed to be a practical work with direct application to instructors and practitioners in the justice field, and to their students. It is written and designed to be a resource in the classroom—both to the teacher delivering an ethics course, and most importantly to you the student. My hope is that this book will become a *working* text; that it will become well thumbed by the end of the educational experience for which it is being used.

Much of the content of this work is characterized by an approach that I can only describe as deeply personal, and for this I beg the reader's indulgence. My experience gained as a correctional worker and institution administrator before I became a college professor serves as a foundation for much of what is discussed. And many of the ideas in the book spring from 20 years of trial and error on my part as a college professor working in a college justice department, as I have attempted to address ethical issues in my own ethics courses. It has been both a privilege and a challenge to work with students over the years, and they have taught me much. The book's genesis and content have also been influenced by the experience I have gained in conducting ethics seminars with about 3,000 participants from various justice departments in Canada, Australia, and the United States. As with my students, these participants have also contributed greatly to my learning. The work also has its roots in the research I completed for a Ph.D. dissertation at the University of Toronto.

Briefly, the dissertation addressed ethics teaching in the justice field; it also examined the potential effects of officer subculture on the ability of individual officers to make independent moral decisions in their (justice) workplace. The dissertation employed a qualitative research methodology, which means that much of the data was gathered through in-depth interviews with practitioners from the justice field. Most, but not all, of these interviews were with correctional officers and administrators. Disappointingly, I experienced some difficulty obtaining cooperation from police forces for the purposes of interviewing their staff about this sensitive dissertation topic. Since completing my dissertation, however, I have had access to a number of police personnel who were prepared to be interviewed for this book.

Reputable Conduct, then, is clearly colored by my own personal experience as a worker and administrator in corrections, by what I have learned from students and participants in my ethics courses and seminars, and from the research for my Ph.D. dissertation.

The Design of the Text

Though designed to be thought provoking, this text is not designed to condemn or accuse or point a finger. The roles of the correctional officer and the police officer are often demanding and extraordinarily complex ones. It would be improper, therefore, for this text, or any similar text for that matter, to peddle a particular point of view with respect to the complex ethical issues that often emerge from this kind of work. There may be times when you feel, however, because of the nature of the challenges presented and the questions asked, that there is a hidden agenda; that is, that there is a preferred way of thinking about a given issue and that you are being led in that direction. Because students can sometimes feel this way, there is a discussion in Chapter 4 about what I call the "lurking dangers of indoctrination," or the perception by students of indoctrination by the teacher. The intention of this book is not to lead you in any particular direction or to tell you what you should do in a given situation. The intention is to prod, push, and poke you into thinking more deeply about what, it would seem to me, are important ethical issues and moral dilemmas facing those of us involved in the justice field.

If the discussions and dilemmas are on occasion posed in a manner that causes you to feel that you are being judged, this is because they are designed to be searching and challenging. It is, of course, not my intention to stand in judgment over you, the reader. On many occasions, I have personally confronted similar dilemmas in my own career as a correctional officer and as an institution administrator, and only a foolish person would claim infallibility in dealing with such dilemmas. Indeed, I am still not sure, many years later, if I did the right thing in some of the situations I faced. Because of my own struggles, and the struggles of others who have conveyed their stories to me, my approach, I hope, is one characterized by compassion. But this does not mean that we should shy away from asking each other some hard questions about our work. As is the case with almost any field of ethical inquiry, there are many more questions than there are answers, so throughout the text we will be stopping, under the heading Reflections, to ask some of these questions.

Perhaps an appropriate first challenge might be for you to debate whether talking about moral dilemmas facing us in our daily work is a worthwhile activity.

Reflections

Might we learn something from considering alternative perspectives of colleagues, subordinates, or superiors?

Is there anyone who has a monopoly on wisdom with respect to the dilemmas we face daily?

Should we go it alone with these dilemmas; that is, work out our own solutions in isolation from our colleagues?

Should we find out what our peers think about specific issues and follow the consensus?

Where does talking about moral dilemmas and their resolution rate as a teaching priority? What about moral, as well as legal, personal accountability?

Does moral accountability matter enough to us to warrant serious discussion about serious issues?

Can this difficult job we do cause our sense of what is right or wrong to become jaded?

Can our moral sense be changed or sensitized or perhaps reawakened?

These are, in my view, examples of important questions deserving serious attention. Perhaps you will consider this point of view itself worthy of discussion. If you disagree with me, then you should be able to give reasons for your disagreement. In any event, the above examples are the sort of questions that will be asked from time to time throughout the book. It would be an interesting exercise for you—at the *end* of the book and your course—to reflect on whether you would make any changes in the way you earlier attempted to answer the questions.

Is There a Link Between Ethics Education and Moral Behavior?

This text aspires to be grist for the discussion mill because I believe that conversation about moral dilemmas and their resolutions has the potential to heighten awareness of personal moral responsibility. Derek Bok (1988), while president of Harvard University, wrote a thought-provoking article called "Can Higher Education Foster Higher Morals?" Bok says:

> Skeptics will reply that courses in moral reasoning have no effect on behavior, but this criticism seems overdrawn. To be sure, no instruction can suffice to turn a scoundrel into a virtuous human being. But most young people arrive at the university with decent instincts and a genuine concern for others. For them, courses that foster an ability to detect ethical issues more perceptively, to think about them more carefully, to understand more clearly the reasons for acting morally seem likely not only to train their minds but to have a positive effect on their behavior as well. (p. 7)

Of course, Bok's notion that most young people arrive at university with decent instincts and a genuine concern for others would apply not only to those of you seeking higher education, but equally to those of you in police and corrections training facilities as well.

I tend to agree with Bok. We have to be realistic about the possible effects of ethics courses, but the potential is there, I believe, for informed and critical reflection and thought to translate into a greater desire to do what one considers to be the right thing. Many participants who have attended my ethics seminars have said things to me that seem to support this idea. Talking about ethics, about what is right and wrong, certainly raises one's level of consciousness about important issues. And being more aware, more thoughtful, about the dilemmas we face can sometimes translate into more courageous decision makeup and actions. But what I think about this issue is not the important thing. What matters most is what you think.

Reflections

What do you think of Bok's point of view?

Do you believe that thinking more clearly about ethical issues and moral dilemmas has any potential for helping you do what you consider to be the right thing?

The Place of Courage in Ethics Education

Taking time out in a course like this to think about ethical issues may, as I have indicated, have the potential to foster moral courage; that is, having worked out what we should do in a given situation, we may have the fortitude to do it, irrespective of the forces pulling us in other directions.

But the first sign of courage, perhaps, is for each of us to be willing to face up to the hard questions posed earlier, and others like them in this text. To risk the intellectual, psychological and emotional discomfort that will invariably result from such honest inquiry is a bold step. It can be disturbing to find that the result of such inquiry is simply, on occasion, little more than the replacing of answers we thought we had with troubling questions concerning those "answers." It can be equally disturbing to be confronted with the thought that if we are to be honest with ourselves, our particular way of looking at something, or a particular value we have held for some time, may need revision—if not complete rethinking.

Students often talk about being somewhat scared of confronting their biases and prejudices. The good news is that those who show the willingness and the courage to submit their values and ideas to scrutiny often feel they are stronger for the experience. One of my students, referring to his personal growth, wrote at the end of an ethics course: "Not only does this course challenge your existing values and norms, I found myself asking why do I think like this? and what causes me to think like this? . . . I think if you are comfortable with this self-examination, that it opens up doors for self-growth." Another wrote: "We were afraid and we grew. . . . I have realized that being afraid is normal." And another: "This has been a challenging class. While I have often left the room with a headache it has certainly made up for it in other ways. True morals are those that can be challenged openly and objectively. If a person can do that and still believe in that moral, then it is certainly a moral that should be believed in."

Reflections

Is being willing to submit your views and values to a significant challenge the kind of activity you consider requires courage?

Or, is it a foolhardy exercise, at best likely only to reduce our general happiness, and at worst likely to confuse us totally?

There are people who would argue, possibly with some justification, that unless we are able and willing to test our views and our values by the fire of honest inquiry and soul-searching, that (1) the values are not truly our own and, therefore, (2) they are not worth having. Many of our views and values, it is argued, are inherited; put on, as it were, as we would a suit of clothes. We have just assumed them, often without paying much attention or thought to how we arrived at them. If this is the case, then perhaps we should call them habits rather than values. What do you think?

Inheriting "Values" in the Justice Field

Some would argue that the tendency toward inheriting views and values described in the Reflections segment above applies *particularly* to those of us who work in the justice field. Some would take it a step further and argue that to resist inheriting or adopting the values of our colleague group is a decision fraught with some risk. These values are what we sometimes call norms. For example, Frederick

Desroches (1986), a sociology professor at a Canadian university, speaking of new recruits to police work, makes the point:

> From association with fellow police officers, and by being sensitive to their attitudes and beliefs, the new recruit soon picks up on the general group's opinion about most police matters. . . . Once a subculture has been formed, novices entering the group are confronted with a "fait accompli"—a set of norms and values to which they are exposed and are expected to conform. (p. 40)

And Grossi and Berg (1991), in the corrections context, make a similar point:

> Again, it may be that in order to attain a satisfactory level of peer support, and reduce conflicting pressures endemic to correctional institutions, correctional officers must compromise their personal values and interests. (p. 80)

Clearly, these are sobering and troubling sentiments. The writers indicate that to survive as a police or correctional officer in terms of early and continuing acceptance by peers, an individual officer would have difficulty operating as an independent person, doing independent thinking about moral dilemmas encountered on the job. Desroches talks about police officers, especially new ones, having to conform to group norms and values. Grossi and Berg, in talking about correctional officers, take the idea one step further and talk about the requirement to compromise personal values and interests in order to continue being accepted by peers.

Later we will discuss this issue of colleague loyalty, or peer pressure, if you like, in greater depth. The issue has to do with *subculture* in policing and corrections, and how it can impact on individual officers.

Reflections

What do you think of the views of these researchers and writers? Do they have any merit? Or, do these two authors have it wrong?

A high-profile case from Canada's armed forces may shed further light on this troubling question.

Subculture in the Military

In 1993, an event unfolded—and I hope I am not being too melodramatic here—that scarred the soul of Canada. Consideration of that event here may help to provide us with a useful context for discussions later in the book. Canada as a country, historically, has taken some pride in the fact that it has contributed to world order and peace by deploying its troops to many parts of the world for peacekeeping duties. Canadian soldiers have been characterized as something approaching angels of mercy and have, more often than not, been honored and accepted by the citizens of the countries in which they were serving.

In 1992–1993, Canadian troops were sent to Somalia on a peacekeeping mission. The soldiers were members of the Canadian Airborne Regiment. A subsequent inquiry has discovered that some of these peacekeeping soldiers conducted themselves in a brutal, racist manner and, in fact, tortured and murdered a 16-year-old Somali youth. One disturbing phenomenon that has emerged from the inquiry is that the military has its own entrenched subculture, and that this subculture is often characterized by some of the things Desroches and Grossi and Berg talk about. One of the chief characteristics is the tendency for recruits to allow themselves to "buy into" the subculture in such a way that their loyalty to colleagues can transcend the responsibility they have to think for themselves and to act according to their own moral sense. Indeed, one military police officer admitted that "his investigations were thwarted by soldiers covering up for each other." While giving evidence at the Somali inquiry, this military police officer added: "We're soldiers first and tradesmen second," later suggesting that if he were the one being questioned by the police, he would probably keep quiet as well (*The Toronto Star,* October 15, 1996, p. A2).

In 1993, on the CBC radio program *Morningside,* during a discussion about the Somali affair between host Peter Gzowski and a retired lieutenant colonel of the Canadian military, the lieutenant colonel referred to what she described as the "dangerous bonding" that can take place in military units; this bonding, she claimed, effectively inhibits—if not prohibits—individual moral decision making. She asserted—and I am paraphrasing here—that individuals (soldiers in this case) must constantly be reminded that they continue to carry full responsibility for moral decision making irrespective of the prevailing ethos or subculture.

In the field of law enforcement, an excellent example of that "bonding" was witnessed in testimony received by the Mollen Commission, the body assembled to look into a corruption scandal that erupted in the New York City Police Department in the early 1990s. One of those giving testimony was a former officer by the name of Bernard Cawley, a man who had participated in a range of egregious behaviors while on duty, and who had been sentenced to prison for those acts. He was brought before the commission and asked a series of questions. In one of the more breath-taking exchanges during his testimony, Cawley was asked "Weren't you afraid of getting caught?" In formulating his answer Cawley paused for a few moments, and then replied, "No. Who's going to catch us? We're the police."

De-Individuation

It may be useful to contrast the lieutenant colonel's point of view with something psychologists call *de-individuation.* Psychology professor James Ogloff defines de-individuation thus: "There is pressure exerted by the group and people lose the ability to regulate their individual behavior." He adds: "You are not held accountable for what has gone on, so you end up doing things you wouldn't do individually" (*Maclean's,* January 30, 1995, p. 18).

It may be difficult, however, to reconcile the lieutenant colonel's view that individuals continue to carry full responsibility for their actions when, if Ogloff is correct, peer pressure or subcultural constraints can cause individuals to lose the ability to regulate their behavior. In other words, that in a sense they become victims. (This apparent conundrum would make a lively topic for debate in your class. But let's continue the discussion.)

A Confession

In a haunting article entitled "Discretion No Part of Valour" in *The Globe and Mail* (October 1, 1996, p. A6), Eron Main, a former Canadian military officer, refers to the Somalia affair. The article is a confession of sorts. Main writes: "The unofficial military code I lived by involved turning a blind eye to my fellow officers' misdeeds. Ultimately such 'loyalty' led to the Somali inquiry as well as to my retirement." He adds: "We learned the lessons of loyalty a little too well, until we gave it where it was no longer deserved: to the officers who had debased their commissions, and were no longer good and honest." Main, in reflecting on his days as a student at the Royal Military College of Canada, talks about the code of loyalty: "In four years I had encountered many future officers who neglected the college motto—Truth, Duty, Valour—in favour of the unofficial code: Don't Get Caught. I knew of several whom I considered unworthy of their commissions. Of course, I would say nothing. We had been taught from our first day as cadets that we should never, ever, betray our friends, our classmates, our fellow officers."

The U.S. Naval Academy is an historic and tradition-bound institution charged with turning out leaders of the U.S. military. One thing that has always made that school (and others like it) seem to stand above and apart from the rest has been its stated support for, and presumed adherence to, a Code of Honor. That code essentially maintains (and I am paraphrasing here) that "We will not lie, cheat or steal, nor tolerate those who do." Several years ago, when the Naval Academy became embroiled in a series of scandals including academic cheating, drug sales and usage, and auto thefts, the television program *Sixty Minutes* conducted a series of interviews. They learned that in addition to the formal Code of Honor there was another less formal but in many ways more powerful "code" that held "Never bilge your buddies." In the police and corrections arena, that "code" would caution practitioners to never inform on or "rat out" their peers.

These sentiments echo many of the thoughts and comments expressed to me by respondents in my dissertation research. They also echo many of the comments made to faculty in our justice department by students when they return to classes after serving their field practica with police forces or in correctional institutions. Now let's try to relate what we have been discussing to justice students and the college classroom.

Knowing, Yet Not Knowing

One of the more difficult challenges I have had to face in my teaching career was dealing with a group of angry justice students who had decided to punish a classmate. The classmate had expressed dismay to a teacher that a number of students had cheated on an examination. Briefly, the teacher, in handing out an examination paper and after instructing the students to leave the papers upside down on their desks, realized she was two papers short. She told the students to leave the papers upside down and she went back to her office to photocopy some extras. While she was gone it appears there was open season on exchange of information by approximately two-thirds of the students! Several minutes later, the teacher returned to a now quiet classroom and instructed the group to turn over their papers and commence writing the examination. One by one the students finished writing the examination and excused themselves from the classroom. The last student to leave was the one I mentioned above. She said to the teacher, thinking there was no other student within earshot: "I have a problem with your leaving the classroom. It allowed several students an opportunity to cheat, and I don't feel that is fair." One of the other students was waiting outside the door and overheard the comment. To shorten a long story, what happened in the next few days was alarming but may come as little or no surprise to many of you reading this account.

Word got out and the reporting student was treated mercilessly by her formerly friendly classmates. On the second day after the incident, she came into my office in a fragile state to declare she could not put up with the hostility anymore. Even though she was in the final term of a two-year program, she had decided to drop out of school. She had committed the cardinal mistake of betraying her classmates. (Perhaps it was Eron Main fearing this kind of reprisal that caused him and his military colleagues to say nothing about the major indiscretions of classmates and colleagues.)

The most sobering thought of all about this event is that these were the same justice students who, the previous semester when we were discussing the notion of subculture in a class, each protested that there would be little or no chance of their ostracizing a colleague for doing the right thing. As discomforting as the examination incident was, and as painful for the student as it was, it provided a useful vehicle for discussion and learning. The student stayed in college, graduated, and is doing an excellent job in the corrections field.

Reflections

What do you think about values and attitudes that can be picked up by new recruits to the military, to policing, or to corrections?

Should loyalty to colleagues be fostered as an appropriate and, indeed, necessary value? If so, what consequences might inheriting this value have on your ability to make the decisions you—and you alone—feel you must make in the context of your work?

Summary

The primary purpose of this book, as perhaps you can see, is to stimulate you the student to think courageously and deeply about ethical dilemmas. The issue of colleague loyalty above all else is just one of many that seems to be an integral part of being employed in the justice field. To be able and willing to engage in honest, unbridled discussion of issues like this one demands a high degree of maturity and courage. For you to resist the temptation to react defensively will constitute a distinct challenge. Of course, if you consider the ideas and questions in this book in a mature and honest way, and you are still convinced that the way in which you thought about them originally can be justified, then that is fine. The intention is not necessarily to change your thinking, though that may happen on some issues; it is simply to provide you with an opportunity for considering the issues in a deeper way.

The idea that it is often difficult to remain one's own person in the face of subcultural constraints constitutes a central theme of the book.

Another important purpose is to encourage you to work toward some personal resolution regarding some of the issues raised, where such resolution can be justified. To be able to achieve this, of course, some tools might be helpful to assist in the process. The good news is that some are available and they will be discussed in detail in Chapter 8. Opportunities will be provided for you to practice using these tools as you struggle to reach some conclusions regarding the dilemmas.

The dilemmas offered for discussion are not hypothetical. It is my hope that they will touch a chord in you because of their practical significance. They are real. They have been experienced and recounted to me either by my students while on their field practica, or by officers and administrators of police forces and correctional institutions in Canada, the United States, and Australia. Some of the specific details I have changed in the interests of ensuring that no police force, correctional institution, or individual is identifiable from the information given. Some of the dilemmas are ones more likely to be experienced by police and correctional institution administrators, and others by frontline officers. I would encourage you to "have a go" at thinking your way through both the administrator and officer dilemmas. In this way, you may become more sensitized to the dilemmas others can experience.

Reflections

Should leaders in policing or corrections encourage subordinate officers to think for themselves?

How do you feel about this idea of thinking for yourself, especially at times when your colleagues may be suggesting strongly that you not do so?

One final note. The fact that the situations included for discussion are *dilemmas* indicates that they probably beg no hard and fast solutions. Indeed, there may well be no one right answer. What we can do is to make a choice that is the best available, while acknowledging that the perfect solution does not exist.

Before we get to the point in the text where these dilemmas are posed, however, you will be aware that chapters 3 and 4 address important topics which, I hope, will become subjects of debate and discussion almost before anything else is discussed in your class. Chapter 3 addresses The Ethics of Teaching Ethics in Justice Programs and Chapter 4, as has been noted, addresses The Role of the Ethics Educator: The Lurking Dangers of Indoctrination. In my experience, it is unusual to encourage students, either in a book like this one or in an ethics course, to participate in a dialogue about these issues. However, discussion of these topics with your instructor will probably be helpful in setting the climate and tone of your classroom experience. It may also help in establishing some parameters.

More important perhaps, it may save you and your instructor some of the pain and anguish my students and I have experienced over the years caused by fuzzy role expectations, unclear teacher intentions, and miscommunication regarding the purpose of the ethics course.

Ideas

Man is not to blame for what he is. He didn't make himself. He has no control over himself.

—Mark Twain, novelist

Nature is strong and she is pitiless. She works in her own mysterious way, and we are victims. We have not much to do with it ourselves. Nature takes the job in hand, and we play our parts.

—Clarence Darrow, American defense counsel

We are responsible. . . . It is not true that everything that happens to us is like "being struck down by a dreaded disease."

—Sidney Hook, philosopher

He who lets the world, or his own portion of it, choose his plan of life for him has no need of any other faculty than the ape-like one of imitation. He who chooses his plan for himself uses all his faculties.

—John Stuart Mill, philosopher

Man has free choice, or otherwise counsels, exhortations, commands, prohibitions, rewards and punishments would be in vain.

—St. Thomas Aquinas, theologian

Give me a dozen healthy infants and I'll guarantee to take any one at random and train him to become any type of specialist I might select—doctor, lawyer, even beggar-man and thief, regardless of his talents, penchants, tendencies, abilities.

—John B. Watson, behaviorist

Chapter Objectives

After reading this chapter, you should be able to:

- *Explain the idea of moral philosophy.*
- *Explain the concept of philosophizing.*
- *Explain the potential dangers of gullibility and mindlessness.*
- *Explain Dr. Martin Luther King Jr.'s ideas about good laws and bad laws.*
- *Explain the difference between morals and ethics.*
- *Explain the term* moral dilemma.
- *Explain what we might mean by the term* a wise person.
- *Define libertarianism and determinism.*
- *Explain utilitarianism and identify two major writers who developed the concept.*
- *Explain the Principle of Equal Respect.*
- *List and explain some ideas about justice developed by Plato, Socrates, Aristotle, Hobbes, and Sartre.*
- *Explain Plato's idea that how we behave is closely linked to the thinking we do about our behavior.*
- *Explain Socrates' ideas about teaching (called the Socratic method).*
- *Explain Aristotle's contribution to philosophical thought in terms of his ideas about logic.*
- *Explain what Hobbes meant by Social Contract.*
- *Explain Sartre's view that "man is condemned to be free."*

Introduction

This is a foundation chapter, so to speak, for the rest of the book. It provides some ideas that a few selected thinkers and writers have had about society in general and justice in particular. Try to keep these ideas and concepts in mind as you move through the book. Make whatever connections you can and never hesitate to refer back to the ideas discussed here because they may help to shed some light on dilemmas raised later.

It's always a pleasure for teachers, and I very much include myself here, when they see their students making connections with ideas they have previously learned. It means (1) that at least some of the things they learned previously have stayed with them (!) and (2) they are not locked away in some kind of compartment. The best students are the ones who can relate ideas and concepts from one class to another and from one course to another. Let me encourage you to engage in the kind of thinking that draws from any and all of your educational experience(s).

Let me further suggest to you that learning that is done in separated, isolated, digestible pieces for later regurgitation in tests or exams is, in all probability, not the best kind of learning because it essentially relies on remembering, telling, and forgetting. Do yourselves and your teachers a favor; look at the ideas and discussion in this chapter as part of an integrated whole, as something to be referred back to as you progress through the book, not as an end in itself to be learned and then moved on from.

"Keep It Simple, Stupid," they said. Now there's a challenge.

Before your eyes glaze over and your senses become numbed by the prospect of extreme boredom, hang in there! I am going to attempt to make this background information as interesting and as relevant as I can.

First, we address some commonly recurring questions: What is moral philosophy? What does it mean to philosophize? What's the difference between morals and ethics? What is the nature of a moral dilemma? What is a wise person? Second, together we discuss some ideas that certain important people have had about moral philosophy. We do so in a *context* that attempts to make these ideas come to life. Third, we examine some ideas of just a few of the major players in philosophical thought. We take a look at three ancient Greek philosophers, Plato, Socrates, and Aristotle; one seventeenth-century philosopher, Thomas Hobbes; and one contemporary philosopher, Jean-Paul Sartre. Of the many big contributors to philosophical thought I have chosen these thinkers in particular because of some of the observations they had to make about the human condition in general and justice in particular.

One of the biggest challenges for any writer of a chapter like this is to convey the material in a meaningful and relevant way. The risk, of course, as I attempt to do so is that I will do a disservice to those very thoughtful, clever, wise, and insightful people. I hope they will forgive me (all of them will have to do so from their graves!) for any injustice I commit. Any distortion of their complex ideas as they are examined through the prism of my attempt to be interesting and relevant, I hope will be forgiven. But the clear message from those requesting the inclusion of this chapter is that, in essence, I should keep it simple. After reading this section and the chapter, you will be the judge as to whether or not I have been successful.

Common Questions About Morals and Ethics

What Is Moral Philosophy?

The word *philosophy* simply means "the love of wisdom." The word has its roots in two Greek words: *philein,* which means "to love," and *sophia,* which means "wisdom." The purpose of philosophy is to explore important matters concerning the world and the universe and those of us who live in it. Moral philosophy or ethics is one of the several different philosophical disciplines that searches for knowledge and meaning in matters of human behavior and morality. Philosophers engage in an ongoing search for truth and understanding, and they do so more by asking important questions than by attempting to find and provide prescriptive answers to life's mysteries. So you would never find a moral philosopher, or any other kind of philosopher for that matter, who would feel at all comfortable telling you what you should do in a given situation or how you as an individual should behave. That search for truth and for a sense of what may be right or wrong with respect to a certain action is your search. And the more you know about moral philosophy and what various thinkers have said about important issues, the more skilled you will become at thinking through your own situations.

What Is "Philosophizing"?

When you think about important questions like the rights of animals, the rights of persons in our society who are mentally challenged, abortion, capital punishment, social injustice, the meaning of life, and so on, you are philosophizing. You are searching for truth and meaning. The only difference between you and philosophers *per se* is that they are probably philosophers by profession, that is, they probably write about it and teach it. They will probably be more skilled because they have spent a lifetime studying the ideas of many great minds and have been trained in the rules of logic, thinking, and argument. So they will almost certainly be more skilled than the rest of us, but that does not mean that when we think deeply about important issues that we are not philosophizing. So there is no magic to it, and we should never feel excluded from the process.

What is important is that we do our own thinking! Merely choosing to go along with the crowd, or merely accepting the status quo, is, I would suggest, to do ourselves a disservice. Not only that but it's also dangerous. We can be led up garden paths and end up deeply regretting our gullibility.

I remember many years ago in Britain how a very credible and well-respected news program called "Panorama," hosted by a very reputable and trusted broadcaster, hoodwinked almost the entire nation one April Fool's Day. The program aired a very serious-sounding documentary on the spaghetti trees of northern Italy. The film footage included scenes of many acres of cultivated trees bearing their spaghetti harvest! The frightening thing is that in this case people were prepared to suspend their own knowledge and beliefs because the information was being conveyed in a serious manner by what they considered to be a very credible and trustworthy source.

Think of the potential consequences of such mindlessness and gullibility. Perhaps it is because of this kind of phenomenon that we had a Nazi Germany or, on a more personal scale, why we sometimes are prepared to perhaps be too accepting of the influential views of teachers or of books. We owe it to ourselves to develop a critical mind and refuse to accept things at face value. (You might like at this point to take a look at the section on critical thinking in Chapter 8.)

So enjoy yourselves as you philosophize about the contents of this book or about any other important issues. You have as much right to do so as anyone else.

Here's something for you to get your teeth into for starters. Many of you reading this book probably already work in the justice system upholding the law. Many others of you may shortly be seeking employment in the field. What motivates the interest in this field for many of you is your respect for law and order. Right? Well, Martin Luther King, Jr., said there are bad laws and good laws. The laws that segregated blacks in the southern United States he considered to be bad laws because they discriminated between human beings simply on the basis of skin color. He argued that it is ethical to obey good laws and unethical to obey bad laws—that is, bad laws should be disobeyed. That would mean civil disobedience. It's an interesting viewpoint and one that we should all seriously consider prior to embarking on our careers in the justice system.

What Is the Difference Between Morals and Ethics?

This is a commonly asked question. The term *ethics* comes from the ancient Greek word *ethos,* which means "character." There is really little or no mystery to the difference between morals and ethics. Ethics, as I have said earlier, refers to the study of morals and morality, about what's good or bad, right or wrong. It refers to the practice of thinking philosophically about morality. When you read and think about the dilemmas in Chapter 8, you will be engaging in ethics or moral philosophy. Of course, you do not have to do your own thinking on these issues and others like them; you can simply choose to accept the majority opinion and go with that. But even that decision, if you think about it for a moment, is a moral decision. In an important sense you have made a decision to relinquish your autonomy and buy into, perhaps unthinkingly, the wishes of the group or the way society generally feels about an issue.

Morals, on the other hand, simply refers to human behavior. Morals have to do with how individuals relate to each other and to their environment. Ethics is the study of those morals. The word *moral* comes from the Latin word *mores,* which refers to social habits or behaviors.

Having understood these distinctions, you should be aware that modern philosophers and folk in general tend to use these two terms interchangeably. We talk about a person's ethics as we would talk about his or her morals and vice versa. So do not be overly concerned about the fine differences between the two terms.

What Is a Moral Dilemma?

In Chapter 9, you will be presented with a number of moral dilemmas. Indeed, you will find them lurking in all of the chapters! So it is pretty important that we have some understanding of what a moral dilemma is. We know we all experience them, but what are they? What constitutes a moral dilemma? Are there some characteristics that all moral dilemmas have in common?

A moral dilemma occurs when we are faced with a very tricky decision about a complex problem. Generally, the problem is one that does not have a self-evident answer in terms of what we should do about it. We become puzzled and worried because we don't know what we should do in the situation. If we decide to do one thing, there is a problem, and if we decide to do something else, there is also a problem. In other words, there may not seem to be any one right answer.

Let's consider the example of whether or not you should tell your best friend about a one-night stand you know her fiancé recently had. The husband-to-be is also a very good friend of yours, and he knows you know of his indiscretion. He begs you not to say anything, that it was a stupid mistake, a spur of the moment thing at a weak moment, that he loves his partner and wants to marry her and spend the rest of his life with her.

Now for some of you this situation may not constitute a dilemma. You may feel strongly that he is a dirty rat and that your friend has a right to the information that he cheated on her. Others of you may feel equally strongly that what happened is none of your business and for you two maxims apply: (1) It's always better to let sleeping dogs lie and (2) what a person doesn't know can't hurt him or her.

For others, however, the solution may not come quite as readily. Many confronted by this situation will agonize, and lose sleep even, over what they should do. The feeling that many have when confronted by such a dilemma is that they are damned if they do and damned if they don't. In essence, they feel the situation is a no-win situation.

There are at least two lessons here. First, if you and I are faced by the same dilemma, it may be the case that only one of us will feel the need to agonize over it. So what strikes you as a dilemma may not strike me in the same way and vice versa. We should not judge each other, because we clearly come at the problem from two entirely different perspectives and life experiences. For example, if my partner has cheated on me in the past I might come to a quicker solution about what I should do! Second, the right solution for you may not be the right one for me. If that is so, I should try to resist the temptation to judge you based on your decision, nor should I attempt to influence your decision.

For this reason, we've got to be really careful at moments like this. When a friend approaches us and confides in us about the nature of the dilemma he or she is facing, we would probably be wise not to be drawn into giving that person advice.

The point of this book, however, is that we can learn to philosophize better, to think more clearly about dilemmas that confront us. Hopefully, as you read the book you will learn about skills, ideas, and tools that can help you in your decision making. These tools are discussed in Chapter 8 and if you take a close look at them you may find one or two of them particularly appealing as you seek to work out your own dilemmas.

Something we really should be wary of is responding emotionally to a difficult problem. Although we cannot entirely discount our emotions, of course, we would be well served to become more skillful in *thinking through* these issues. The tools discussed in Chapter 8 may serve to provide a framework for that thinking.

What Is a Wise Person?

Of course, if I thought I was able to offer you a definitive phrase in response to this question I would only be showing how unwise a person I am! For anyone to claim to know the answer to such a complex question should make us wonder about the person making the claim.

Perhaps the most we can hope for is to attempt to identify some qualities a wise person might have and, on the other hand, suggest some things that a wise person is not.

I may have it all wrong, but I often tell my students that an expert in the justice field is someone who knows best what doesn't work. (And, as you might imagine, many of my colleagues feel this is far too pessimistic a view to take. Do you agree?) Of course, on the contrary, we often hear of different individuals making absolute claims about what will work: "We're too soft on offenders"; "Lock 'em up and throw away the key"; "The criminal justice system is too harsh."

Now I don't want to sound too defeatist here but at one point or another virtually all of the "new" methods about how to deal with offenders have been tried—and found wanting. So anyone claiming to know answers to very complex problems may be the very antithesis of what we consider to be a wise person.

The great philosopher Socrates was brought to trial in Athens in 399 B.C. for supposedly speaking out against the gods of his day and for corrupting young minds. He was put to death by being forced to drink a cup of hemlock, a poison. His unpopularity had its roots primarily in his ability to expose those who claimed to be wise but were not. In his defense he said something quite significant. He had been talking about a politician whom people had thought wise but who had been exposed by Socrates to be a fraud. He said in his defense:

> Well I am certainly wiser than this man. It is only too likely that neither of us has any knowledge to boast of ; but he thinks he knows something which he does not know, whereas I am quite conscious of my ignorance. At any rate it seems that I am wiser than he is to this small extent, that I do not think that I know what I do not know.

A wise person, then, is probably someone who would seldom if ever claim to "know" much about anything, particularly in an absolute sense. He would be a person who would be cautious about making any kind of absolute claim. He would be thoughtful and well read about any topic before venturing an opinion on that topic. He would be open minded and not afraid to look at evidence that contradicts or competes with his current views. He would have a critical mind and be always on his guard against unthinking acceptance of ideas that had not been thought through. He would certainly resist any pressure or temptation to hold what we sometimes call a "blind belief."

Take care as you wrestle with the dilemmas presented in this book that you don't come on too strong about what the "right" or "wrong" response is. Try as hard as you can (and I know from personal experience it is terribly difficult) to be humble and to be conditional and measured in your thinking.

Of course, to seek to become a wise person, as an end in itself, is no doubt doomed from the outset. We can certainly attempt to conduct ourselves in such a way that we aspire to some of the qualities of a wise person. But to seek to become a wise person as an end in itself will almost certainly result in frustration and disappointment. Wisdom may be a bit like greatness in this regard. Generally, truly great people do not seek to become great; they have greatness thrust upon them. It is as though greatness chooses them, not the other way round. And so it might be with wisdom. Seek it for its own sake and it almost certainly will be elusive. Seek to be honest, open minded, cautious about the truth, and humble as you search for knowledge and meaning, and wisdom may find you.

Reflections

When is learning not learning?

What do you think of the idea that when we are thinking deeply about important issues, we are philosophizing?

What is your definition of wisdom? Of a wise person?

What do you think of what Socrates said in his defense about knowledge?

Most of us, at times, and I very much include myself here, have come on strong about having "answers" to very complex problems. Why do you think it is that we set ourselves up in this way? What drives our need to set ourselves up as experts?

There is an expression I have heard, and perhaps you have heard it too, that goes something like this: "The more we know, the more we don't know." What might this mean?

Ideas in a Context

Perhaps the best way to examine these ideas, as I have already hinted, is to look at them in *context.* That is, we will take an issue that is relevant to the criminal justice field and examine it from the perspective of some of the different ideas and theories that have been written about. Of course, this examination, as you shall see, will be a cursory one—it will be neither in-depth nor exhaustive. We will touch on only *some* of the ideas that are out there. Keep in mind that you may feel that the context is introduced in a provocative way. The approach is deliberate, but it should in no way be construed as my peddling any particular point of view.

The Context

Here's the context. Every society has its criminal element. It is a practice in most societies, particularly ours, to deprive offenders of their freedom when they do something wrong (some of us may think not for long enough). The purpose of locking offenders away, we say, is to give them time for reflection, to provide a safeguard for the rest of us in society while they reflect, and to punish them for their wrongdoing. Underlying our system of justice is the idea that people are free moral agents and that if they knowingly commit an unlawful act, they will be punished accordingly.

So let's examine the practice of locking aberrant members of our society away, and how this practice squares with some of the philosophical ideas. Keep in mind that we will just touch on these ideas as a way of getting the thinking juices going, as it were. None of the arguments or ideas are meant to be definitive in any way.

Libertarianism

Libertarians make the assumption that we all have free will to make our personal choices and that based on the choices we make, we should be praised or blamed. If we do good things, we should be praised, and if we do bad things, we should shoulder responsibility for those acts and take whatever consequences that result. William James (1842–1910), who was both a philosopher and psychologist, was one of the main exponents of this point of view.

If we think about it for a moment, our system of law is predicated on this very basic assumption that we are what philosophers sometimes call *free moral agents.* This means that we should consider our options very carefully before we act because we will be held accountable for our actions. When we do something wrong and we are charged with an offense, the court will establish whether we are to blame or not, and frequently the court's decision will rest on whether or not it can be shown that we acted with criminal intent. The notions of *mens rea* and *actus reus* loom large in criminal law. *Mens rea* refers to the idea that we acted purposefully and with evil intention. *Actus reus* refers to that which is forbidden by law.

Put simply, the theory goes something like this:

Fred broke the law.

Fred knew that he was breaking the law.

Fred could have acted otherwise.

Because of the choice he made, Fred deserves blame and punishment.

It is just and fair to punish Fred because he freely chose to do a bad act.

If this kind of thinking is correct, and it may well be, then our system of law would appear to be just and fair. The maxim "let the punishment fit the crime" appears to be a reasonable one if we are, indeed, free to make individual choices. Potential offenders should know that society is not going to respond too kindly when they violate the freedoms of the rest of us.

Certainly, it is only by thinking in this way (that is, that we are free moral agents) that we can make much sense of our lives, our accomplishments, our system of rewards and punishments, of praise and blame, of accountability, and of personal responsibility.

Reflections

What do you think? Do you think we are free to make our choices and that we should be held accountable for those choices?

What do we mean when we say people are responsible for their own actions?

There is, however, another way of looking at this issue of whether we do, indeed, have free will. There is another school of thought called *determinism.*

Determinism

Rather surprisingly perhaps, novelist Mark Twain of *Huckleberry Finn* fame was a leading advocate of the determinist point of view. Here's something he wrote:

> Man is not to blame for what he is. He didn't make himself. He has no control over himself. All the control is vested in his temperament—which he did not create—and in the circumstances which hedge him round from the cradle to the grave and which he did not devise. . . . He is purely a piece of automated mechanism as is a watch, and can no more dictate or influence his actions than can the watch. He is a subject for pity not blame. . . .

For Twain, then, we are victims of our circumstances. We are shaped by heredity and by the particular and peculiar environment we experience. We have no control over the process. To all intents and purposes, human beings are like dried leaves being blown here and there by a winter wind.

We are subject essentially to the continuum of cause and effect. Each one of our behaviors is conditioned or caused by something else. The bottom line is that as human beings we can become no other than what we become. Sometimes we refer to this kind of thinking as *fatalism*, that what will be will be and there's not much we can do about it. Also, those of you interested in theology will no doubt be familiar with the term *predestination*. This concept, derived principally from the writings of John Calvin the theologian, speaks to the idea that God has determined for each one of us ahead of time what our fate will be in this life and in any life following this one.

Now if the determinists are correct, there may be far-reaching implications for our justice system. For example, we may well ask ourselves whether it is fair and just to punish someone for committing a murder when, in fact, she could have acted no differently. The thinking is that the act, in a sense, chooses the person, not the other way round. The person becomes who she is, including all of her choices, because of her quirky birth and life experience.

Let's have a go at applying this way of thinking to the justice system. Have you ever considered how our courts sometimes work? Would you agree that there might be some substance to the idea that an accused person, dressed very well and with a reputable lawyer at his side, may experience a more favorable verdict than a person who is unkempt, has few if any social graces, and is poorly represented? Now I know that I am stepping into shark-infested territory here, but I am doing so, as I said earlier, to get the intellectual juices going. I am being deliberately provocative, so it is important you understand that this discussion is not to be construed as an attempt to influence you.

Let me tell you about an experience I had many years ago when I was working as a parole officer. I was on court duty. Often (and I have to be careful here not to overgeneralize or to become guilty of gross stereotyping) a criminal court is populated by what we call the underclass in our society. They are the "have-nots"—the poor, the downtrodden, the institutionalized—the marginalized members of our society. From time to time, however, sitting in the courtroom awaiting their turn before the judge are what appear to be more privileged members of our society—professional folk, people who are well groomed and attractive.

On the particular day in question, I looked around the courtroom and noticed two very attractive and immaculately turned-out individuals, a man and a woman, husband and wife. It turned out that they were both employees of a major airline and he was before the court on a charge of shoplifting. The charge was that he was in one of the local grocery stores and stole $170 worth of groceries. Apparently, he wore his airline overcoat and stuffed the numerous inside pockets with pork chops, butter, and candies and the list went on. When his case came up, his lawyer told the court what a law-abiding, productive member of society his client was, but that due to tremendous job pressures he had found himself acting out of character on the day in question. He didn't know why he did what he did, he was deeply remorseful, and his professional and lovely wife was in the court that day to show her support for her husband. The prosecutor responded in a conciliatory manner and the accused was given a conditional discharge.

By contrast, the very next person to appear before the judge was a scruffy, unkempt individual, with a previous criminal record. He had declined representation, he said; yes, he had been in the grocery store (the same one as the airline person). He was hungry, so he took a loaf of bread. He told the judge that he was pleading guilty and that he should be sentenced so that he could get it over with.

The judge decided that this request amounted to insubordination and disrespect to the court and ordered the accused remanded in custody pending the preparation of a presentence report. This individual spent the following six weeks in custody.

Let's think of this case from the perspective of the determinist's viewpoint. In one sense, logically, criminal courts ought to look more favorably rather than less favorably on society's underclass. It might be argued that they know no better. Their conditioning, often, has been to survive by taking from the "haves" in society. And, often, they feel little or no remorse about doing so—it is the way they have learned to survive.

On the other hand, the airline crew member presumably had a much better life experience with parents who loved him, a wife who cared for him, an employer who valued his services. His life experience was one of good schooling, regular church attendance, and community leadership.

Logically, it could be argued that the courts should come down much harder on those of us who ought to know better because of the way good fortune has smiled on us in terms of our genetic makeup, and the experience of enjoying a supportive, constructive, nurturing environment coupled with the benefit of good modeling in terms of respect for the law and the rights of others. The courts, however, may work the other way, as they did on the day in question.

Let me give you another example, only this time, rather than looking at the courts, we will consider how police officers might sometimes be inclined to do their jobs. This is confession time! Two days ago my wife and I were in a hurry and I was driving over the speed limit. I saw the cruiser—but too late—my goose was cooked. The lights went on, he followed me and pulled me over. Was I aware that I was driving at 81 mph in a 65 mph zone? Yes. Was there a reason for my speeding? Not really, but we were running late. The officer asked to see my driver's license and insurance. He disappeared into the cruiser, reappeared five minutes later, and told me that I could have been facing a fine of $150, but that he was going to give me a break. The fine would only be $35 with no loss of points.

Notwithstanding my gratitude for the "break," I have to wonder what the response would have been had my vehicle been old and had I been scruffy? Would I still have received the benefit of the break? You judge. Interestingly, about half an hour later as we continued our journey we saw the lights of two cruisers ahead of us. A truck had been pulled over and the police officers were questioning the driver on the side of the highway. He did not present the most attractive of images! "Oh, oh," said my wife, "that poor guy is going to get the book thrown at him." Now maybe he received the benefit of a break similar to mine, but I rather doubt it.

Reflections

Think about your life circumstances. To what extent do you think you are primarily responsible for them?

Think about what you had for breakfast this morning. Think about the clothes you are wearing today. What factors, if any, played a part in the "choices" you made?

In light of the foregoing discussion is our justice system a fair one? If you think it is, why do you think so? Think of some reasons. If not, why not?

What if the purpose of our having laws were to prevent the "have-nots" from taking from the "haves"? That is, what if our laws were there essentially to protect the privileged in our society? What then?

Is it fair to hold as morally responsible for their actions individuals who don't know what moral responsibility is?

What do you think my response to the "break" on the speeding ticket should have been? Perhaps a more honorable response on my part might have been "Officer, you caught me fair and square, I do not deserve any break, give me the ticket and I'll send off the check for $150."

Utilitarianism

Reference is made to this principle in Chapter 8, Tools for Moral Decision Making. The principle was formulated mainly by John Stuart Mill (1806–1873) and Jeremy Bentham (1748–1832). Utilitarianism is also known as the principle of benefit maximization. Stated simply, the principle contends that a decision made or an action taken by an individual or a state is justified or not justified by its consequences. If, as a result of a decision and/or action, the result is less pain, or greater good, or the world generally becomes a better place, then this is sufficient justification for the decision. It is the consequences that matter rather than the inherent rightness or wrongness of a particular decision. So here it could be argued that if my telling a lie results in the world becoming a better place, then that justifies the lie.

Let's examine this principle with respect to our jailing lawbreakers. I guess it could be argued that if our society, generally, is a less dangerous place because of the existence of correctional institutions, then that is a sufficient reason for having them. But what if it could be argued that our society is a less dangerous place—but only *temporarily*? What if it could be argued that, in total, society becomes a *more* dangerous place, because in institutions offenders learn new tricks, and more often than not, their anger and resentment toward society may very well build because of their experience of incarceration and that when they get out, we're all fair game for their next antisocial act because of bitterness?

There are some serious implications here for *why* and *how* we incarcerate. If our methods of incarceration are designed to be repressive and harsh, might they become counterproductive in the long term because most offenders at some point will be released? We may do well to ask what kind of people are going to be let back into our communities?

Reflections

Would you agree that generally the consequences of treating people reasonably are better than those of treating them unreasonably? Think about your own experiences and how you respond. What might be the implications here for policing and corrections?

What is your view on this issue? Do you think that the principle of benefit maximization should be considered as a guide when we think about what kind of justice system we want?

If a repressive justice system can be shown to ultimately lead to a less safe society, would that be sufficient reason to become perhaps less vengeful in the way in which we deal with offenders?

What if, in your view, you think the justice system is too soft on offenders, that there is no real deterrence anymore because of the way the courts are responding too leniently. How would you square this view with the idea that a repressive, punitive system may cut society's nose off to spite its face?

The Principle of Equal Respect

This principle is referred to and discussed in Chapter 8, Tools for Moral Decision Making. Briefly here, this principle purports to be quite central to our system of justice. Put simply, the principle exhorts us to treat each other with equal respect and requires that we should all be considered equal before the law and that purely and simply on account of our being human that we have intrinsic worth. It follows then, if this is the case, that *all* human beings, irrespective of education, class, color, or any other differences, are of equal worth.

The utilitarians argue that we should treat each other well because the consequences of doing so would likely be better. The principle of equal respect, on the other hand, would require of us to treat each other well simply because that is the reasonable and decent thing to do. Individually, it is our being human that buys us a place, as it were, in the equal respect club.

In their book *The Ethics of School Administration*, Kenneth Strike, Emil Haller, and Jonas Soltis (1988) make the point that this principle contains three basic ideas:

1. That we treat people as *ends in themselves rather than means.* This means that we are morally obligated to consider the welfare of all people.
2. That we accept that all humans are *free and rational moral agents.*
3. That no matter how people differ, as moral agents they are of *equal value* and are owed all basic human rights.

If we think about this principle for a moment we can see where the legal notion of due process may have its roots.

Now what might be the implications of the principle of equal respect for the justice system?

Well, let's think for a moment about arrest procedures. Perhaps it would be appropriate, as and when time permits, of course, for officers, when making an arrest, to think about how they would feel if they were the accused. The thinking process does not necessarily have to be a long and laborious one; it can happen in an instant.

What if this person being arrested were my father or mother, my sister or brother? What would I wish their arrest experience to be? Of course, our response would obviously be dictated at least to some degree by the behavior of the person being arrested. But nonetheless even if my dad were being uncooperative when he was being arrested, would I still expect him to be afforded certain courtesies?

What if I am a correctional officer and I am supervising an inmate who has just received a "Dear John" letter. His wife has informed him that she has moved in with his best friend, also an offender, who himself was just released from jail last month. When the inmate reads the letter, he becomes very angry and aggressive. He is banging around in his cell and letting go with all the expletives he can muster. You ask him to quiet down and he tells you to go away in a biological manner of speaking!! The rules require you to book him. What are you going to do? Are you going to respond legalistically? Or are you going to take a moment to find out what's happening, and then, after you have done so, are you going to attempt to put yourself in his skin? What if you were in his shoes? How would you feel? Are you going to respond to him as you would hope other officers would respond to you if you were in the same situation? Or is this not what the principle of equal respect is all about?

A key question we might ask is "Are we all really equal?" Or might there be one law for some, and another law for the rest of us?

Of course, there's much more we could say about the principle of equal respect and its implications for the justice system. What is important now is that you do some of your own thinking about what those implications might be.

Reflections

If we follow this principle to its logical conclusion might it mean that we should respond to everyone, either in the courts, out on the street, or in our institutions, in a similar way irrespective of the circumstances, including age, race, gender, and/or sexual orientation?

If we think there is a place for the use of discretion, for example, in the criminal justice system, how is it possible to square that with the principle of equal respect?

The codes of conduct of many law enforcement agencies talk about applying the law without fear or favor. Sounds like the principle of equal respect doesn't it? Where do you stand on this?

How do you think you respond to society's underclass? They are the people with whom police and corrections personnel work primarily—the poor, the disenfranchised, the minorities. Do you think of each one of them equally because they share your humanity?

Famous Philosophers Speak Out

As I said in the introduction, this is where we are going to look very briefly at some of the big names in philosophical thought. Of course, what follows are very brief references indeed, but I hope they will be enough to whet your appetite and fire a curiosity in you to find out more.

Plato

Plato lived from 428 B.C. to 348 B.C. so he lived to be quite an old man. He is best known for a very famous work called *The Republic* (sometimes known as Plato's *Republic*). As a young man, Plato established a friendship with Socrates, who is the next person we are going to talk about. Socrates was both his mentor and teacher.

The Republic, briefly, is an account of Plato's thoughts on society and how society would be best shaped. It is a dialogue where arguments and points of view are put forward and discussed. Plato is particularly interested in the place of justice in society. And one of the main injustices, as he sees it, is the gap and the conflict between the rich and the poor.

Francis Cornford, a philosopher, commentator, and writer on Plato and *The Republic*, says that the main question to be answered in *The Republic* is: What does *justice* mean, and how can it be realized in human society? The Greek word for *just*, says Cornford, can have many meanings including observant of custom or of duty, righteous, fair, honest, legally right, lawful—what one ought to do. Justice, says Cornford, has to do with the individual's conduct as it affects others. The relevance of this kind of discussion to those preparing for a career in the justice system is fairly obvious. As you wrestle with the dilemmas presented later in the book, it would seem important to include the concept of justice in your thinking.

Plato examines what would constitute a just state and how members of that state would behave in terms of their relationships with one another. What is good is discovered through the use of *reason*. The thinking and knowledgeable individual is the one who will be a good person and do good acts. Bad or immoral acts, on the other hand, are the result of ignorance.

Here is one of Plato's thoughts, among many others, that might have significant relevance for us now and how we do our jobs as peace officers: "Our object [or objective] in the construction of the state is the greatest happiness of the whole, and not that of any one class."

Perhaps the biggest contribution of Plato is that he has set an example for us that we should think very carefully about the kind of society we want to live in, the kind of people we are, and the kinds of behaviors we engage in and how they affect others.

Socrates

Socrates was a philosopher who lived from 469 B.C. to 399 B.C. He was a funny looking fellow, had a much talked about marriage and several children, and for all of his life was as poor as a church mouse. Eventually, he was put to death by being forced to drink a cup of poison. The charges against him were that he spoke out against the gods that the citizens of Athens believed in and that he corrupted the minds of the young. Plato's mind would have been one of the ones that Socrates was accused of corrupting. When Socrates died, Plato said: "This is the end of our comrade, a man, as we would say, of all then living we had ever met, the noblest and the wisest and the most just." Socrates never wrote a word himself; his ideas and his wisdom were captured and written about by Plato.

Even though he died well over a couple of thousand years ago, his thinking is still very influential today. He is famous for developing what we now call the Socratic method of teaching, something you may experience, in fact, with your teacher or instructor. It was certainly an experience that Plato would have enjoyed when he was a young student of Socrates. The Socratic method simply refers to the free and open discussion of ideas, and this, Socrates felt, was the only way to achieve knowledge and enlightenment. In Socrates' day, these discussions would more often than not take place on street corners and, indeed, anywhere where there was a gathering of folk interested in dialoguing about important matters.

One important legacy of Socrates is one very simple but profound thought that he left us: "The unexamined life is not worth living." What he is saying essentially is that if we elect to be dead from the neck up, if we choose not to use our reason or intellect, our lives are going to be much less meaningful as a result. It was important for Socrates, and I would suggest it is equally important for us, to face squarely issues concerning human existence and what they mean. It will take courage, but at least we will feel alive and more complete. And, as a result, we will likely become more knowledgeable.

Becoming more knowledgeable was important for Socrates because he believed very strongly that there was a close link between *knowing* what is the right thing to do and *doing* the right thing, and that thinking about how we should behave leads to better behavior. He also believed there is a close link between behaving morally and being happy. Perhaps, paradoxically, however, he felt strongly that the wisest person is the one who realizes how ignorant he is.

One final thought about Socrates. Both he and his friends were keenly aware that Socrates had been imprisoned on trumped-up charges and that his imprisonment and death sentence was an injustice. His friends, particularly one very close friend Crito, wanted to help Socrates escape. Socrates would not hear of the plan because for him to escape would have been to go against the laws of the city. He said to Crito: "Do you imagine that a city can continue to exist and not be destroyed if the legal judgments in it have no force, but can be nullified and destroyed by individual persons?" Clearly, Socrates was very much against anarchy. His respect for the law was absolute.

Earlier we noted what Dr. Martin Luther King, Jr., had to say about bad laws and good laws. He said that we should respect and obey good laws but that we have a moral obligation to disobey bad or unjust laws. Where do you stand on this issue? Are you with Socrates or with Dr. Martin Luther King, Jr.?

Aristotle

Aristotle was one of Plato's students. He lived from 384 B.C. to 322 B.C. He was a kind of argumentative student who had many disagreements with his teacher. Not that there is anything bad about this, of course. One of the greatest compliments a student can pay his teacher is to feel comfortable enough to engage in open, honest discussion with him.

Aristotle was both a scientist and a philosopher. His science background led him to be very practical and down-to-earth in his thinking. It was because of his background in science that he came to devise what we call formal logic. You may also hear it referred to as Aristotelian logic. *Logic* simply refers to rules of argument or of applying correct reasoning and making valid inferences.

One of the more common examples of formal logic, which you may have seen in other books, is this:

Socrates is a man.

All men are mortal.

Therefore, Socrates is mortal.

We also call this kind of argument a *syllogism*. You can see that the conclusion follows from the premises (the first two statements). If the premises are true, as they are here, then the conclusion will also be true. But if the premises are false, then the conclusion will also be false, even though the argument has a logical form. So in arguing correctly there are two important considerations we should keep in mind: the truth of the premises and the form of the argument.

If you remember what we said about the determinists' point of view, here is how they would frame their argument syllogistically:

Every event has a cause.

Every human choice or action is an event.

Therefore, every human choice or action is caused.

Here, the argument is a sound one with respect to its form. However, we do not know for sure if the premises are true, so we cannot be sure if the conclusion is true.

I remember seeing a film on moral philosophy once, but I can't remember its name so, unfortunately, I can't properly attribute it here. The narrator was talking about good and bad argument, sound and unsound reasoning. He used the following example to show sloppy and unacceptable thinking. He made the point that if I think abortion is wrong because killing is wrong, then there would be something wrong if I subsequently declared I was in favor of war. There is a basic flaw or inconsistency in my argument because the major reason I cite in support of my first viewpoint does not carry over into the second.

Consider your thinking as you go through this book! Are there any inconsistencies in your arguments. Are your thoughts logically connected or are they contradictory?

Now this brief discussion about Aristotle's contribution is not the place for an in-depth discussion of formal logic. This is just a taste. But you might like to take a course in logic sometime. You may find that it will help to clarify your thinking.

Other than this significant contribution of logic, it is important to note that Aristotle was what we call a *synoptic* philosopher. He was a man who had a very broad range of interests. He was interested in all of the sciences and, of course, he was passionate about philosophy. He attempted to integrate all of his interests into his thinking and writing. By so doing, he was attempting to develop what we call a "world view" by integrating empirically known information with rational thought about the world.

Plato, Socrates, and Aristotle were ancient Greek philosophers. We now move just a bit closer to today and look at one of the well-known philosophers of the seventeenth century.

Thomas Hobbes

Hobbes was an English philosopher who lived from 1588 to 1679. I have chosen to include him because his idea about something he called a *social contract* has relevance to how we think about our systems of government and justice as we know them. His main work was called *Leviathan* and in it he develops a system of thought called *ethical egoism*. Ethical egoism, as you might have guessed, refers to the notion that people ought to act in their own interests.

It is Hobbes's thinking on the social contract, however, that is of most importance for us because the social contract refers to the rules of justice, how they came about, and how we should respond to them. The rules of justice came about because people agreed this should be the case; they also agreed that the rules should be obeyed and that a government that enforces them should be obeyed and respected, too. What it all amounts to is that we have all agreed to be subject to the constriction of these rules because it is in our best interests to do so. We give up some freedoms in order to gain a greater measure of freedom. If I obey the rules, some of which will necessarily restrict my freedom, and you do too, we can all live happily ever after, right? If you decide not to obey the rules, then it appears only reasonable that you should be held personally accountable and made to do so.

This contract, for Hobbes, is a kind of glue that will hold a society together. It will bring order to a society. There will be relative order and peace, and we will be relatively safe from each other because we have all bought into the plan. Without the contract we would all be at each other's throats and, essentially, we would be living the sort of brutish existence to which the schoolboys sank in the book (and subsequent film) *Lord of the Flies.*

The social contract then is essentially self-serving. It protects you but, more importantly, it protects *me.*

Let's think about this theory for a moment. If Hobbes is correct in his thinking—and he was generally supported later by John Locke (1632–1704) and Jean-Jacques Rousseau (1712–1778)—the theory makes sense if we all had "stuff" to start with. But what if I don't have anything worth calling my own? Why should I agree or be subject to a social contract? Such a contract would be in the best interests of those who have something worth hanging on to.

Who knows where the truth lies? But it is a conundrum to which we should give some serious thought as people charged with the responsibility of upholding law and order.

Keep in mind that mine is a very simplistic and rudimentary explanation of some of Hobbes's thinking. Let me encourage you to do some exploring and digging on your own.

Finally, we are going to take a look at a contemporary philosopher.

Jean-Paul Sartre

Jean-Paul Sartre was a French philosopher who lived from 1905 to 1980. He was what we call an *existentialist.* Existentialist philosophers write about the human experience, principally about human freedom, and about the importance of the idea that we are responsible for our actions.

Sartre leaves no doubt about his view that we are and must be responsible for our actions when he says:

> Existentialism's first move is to make every man aware of what he is and to make the full responsibility of his existence rest on him.

The view that we do not have free will, for him, is an outrageously false idea. It is a cop-out. You may remember our discussion earlier in the chapter about determinism. For Sartre, determinism is the ultimate cop-out. Instead, he feels "I must carry full and complete responsibility for my choices because nothing dictates them except me." But, for him, there is a downside. The downside is that with freedom comes awesome responsibility—and consequences!

Sartre talks, in fact, of our being "condemned" to be free. It is almost as though if we could avoid the responsibility of freedom, we should. But we can't. Our choices and their consequences affect not only ourselves and those close to us but also conceivably the entire world. This is a thought that ought to haunt us whenever we make any decisions of substance.

For Sartre, there is no escaping this responsibility. As James Christian, in describing Sartre's views, says in *The Art of Wondering:* "There is nothing to help us—because the moment we become conscious of what we are, then we become responsible for everything we are and do."

It is interesting to note that Sartre's views on freedom may have arisen to a great degree from his experience of the Nazi occupation of France in World War II. He wrote of this experience:

> We were never more free than during the German occupation. We had lost all our rights, beginning with the right to talk. Every day we were insulted to our faces and had to take it in silence. . . . And because of all this we were free. Because the Nazi venom seeped into our thoughts, every accurate thought was a conquest. Because an all-powerful police tried to force us to hold our tongues, every word took on the value of a declaration of principles. Because we were hunted down, every one of our gestures had the weight of a solemn commitment.

Of course, as spirited as Sartre's defense is of the idea that we have freedom of choice, that does not mean to say he is correct. As we have noted earlier there may be significant difficulties with this point of view. Happy thinking!

Summary

We have not even scratched the surface of philosophical thought in this chapter. But our hand has been on the surface, and perhaps the next move is up to you to scratch it by exploring further!

Of course, there are many other giants in the field who have contributed much to our understanding of who we are as human beings and of the world in which we live. I hope you will be encouraged to continue reading on your own. But beware: Some of these writers and thinkers can be terribly difficult to read. For many of them you will need your headache pills handy. I remember one friend saying to me that many philosophers operate on this principle: "Why say it simple when you can say it complicated!"

In any event, I hope this discussion has stimulated some thinking of your own. When some of the greatest minds cannot agree on important issues, it is of little wonder that we disagree with each other in our conversations. That is the way it should be. In a healthy debate there should always be room for disagreement and difference of opinion. That is how we learn from each other.

I hope such learning will be your experience as you continue to discuss the ideas in the rest of this book. Do not be slow to refer back to these ideas if you feel they can shed some light on a problem you are thinking about and/or discussing. Remember, that's what good learning is all about.

The Ethics of Teaching Ethics in Justice Programs

In taking the ethics course, I have now established a better understanding of who I am, what I stand for, and for what reasons.

—Journal entry, justice student in ethics course

I have a hard time separating ideas and feelings.

—Journal entry, justice student in ethics course

Feelings from time to time are going to be hurt. We don't want to hear or face the possibility that our opinions may be biased or prejudiced.

—Journal entry, justice student in ethics course

One thing that I don't like about the course [ethics] was that the teacher never really shared his views with us. I'm not sure it is fair to expect the class to share their personal feelings with an instructor who will not engage in any sort of self-disclosure. I would have liked the instructor to allow the class discussions and then maybe take five minutes or so at the end of the class to share his thoughts with us. Kind of like the "Final Thought" on the *Jerry Springer Show* (ha, ha!).

—Journal entry, justice student in ethics course

Chapter Objectives

After reading this chapter, you should be able to:

- *Explain what we mean by* ***teaching ethics.***
- *Explain what the purpose of ethics education might be.*
- *Identify reasons for including ethics education in curricula.*
- *Explain what is meant by* ***advice-giving*** *by a teacher.*
- *Identify in what situations advice-giving may be appropriate.*
- *Explain the collaborative approach to ethics education.*

The Debate over Ethics Education

It may be clear to you from the first chapter that I believe a good case can be made for the wisdom of including ethics courses in criminal justice training programs. This view itself, however, bears some examination, for there are some strong and articulate voices that express reservations about the usefulness of including ethics education in any kind of professional training program or academic course. In the interests of honest inquiry, where all views, values, ideas, and opinions are subject to scrutiny, and in the interests of modeling what is being asked of you the reader, it makes sense to include you in a debate about the usefulness of ethics education in preparation for work in the justice field. Keep in mind that what follows is a cursory glance at what various people have to say about this topic. The intent is to offer something that will make you begin thinking about the whole area of ethics education, with the hope that you will follow up with more reading and research of your own as you seek to make your opinions and views informed ones.

The "Teaching" of Ethics

Perhaps an appropriate starting point to the discussion would be to consider what we mean by the phrase "teaching ethics"? Can ethics be taught?

"Teaching ethics" is an unfortunate term, and one that is likely to lead to considerable misunderstanding. Any teacher who teaches ethics, in the strictest sense of the term, would presumably have to be an ethics expert in the sense that she would have to know the right answer to all kinds of moral dilemmas, including dilemmas faced by you and me as individuals. Clearly, it would be a foolish person who would claim to possess this kind of all-encompassing wisdom. Such an expert would also have to know, presumably, which of the tools discussed in Chapter 8 would invariably be successful in enabling us to make "correct" moral decisions. Few people would be likely to lay claim to this kind of expertise. Indeed, it is quite probable that those coming closest to legitimately being able to make such a claim would be the most humble and self-effacing about their expertise. Have you ever noticed how those who are most wise are often the first to question their wisdom, and those who are not wise are often the first to lay claim to wisdom? So, if there is one thing we can probably agree on in this text, it is that such absolute wisdom is most unlikely to exist. It may be more appropriate to think in terms of a teacher or trainer *facilitating* the ethics educational experience, rather than *teaching* ethics.

Reflections

What do you think about the idea of someone "teaching" you ethics?

John Chandler, professor of ethics, in Martinson's *Discourse on Ethics* (1987) talks about reflecting on his role as an ethics professor and his pondering Socrates's question "Can virtue be taught?" He makes a clear distinction between teaching and learning and makes the point that "surely, morality and ethics can be learned. . . . Whether morality can be taught discursively arouses some skepticism in all of us, but certainly there are institutional arrangements by which values can be learned" (p. 9). These "institutional arrangements" would presumably include classes facilitated, if not taught in the traditional sense, by professors. Chandler does not fully explain what he means. In any event, I think the distinction between teaching and learning is a useful one. So, rather than talking in terms of teachers and trainers "teaching ethics," it may be better to talk in terms of teachers providing opportunities for you and other students to think and learn for *yourselves* about ethical decision making.

One of my students, at the end of an ethics course, wrote in a journal: "What you are doing is helping us understand the practical dimensions of ethics in general and how they relate to specific corrections issues. You are leading the horses (us) to ethical waters. Whether we drink or not is up to us individually."

Another wrote: "I can honestly say that I was quite surprised that you had not overtly attempted (as far as I could tell) to influence our thinking. I thought you would attempt to try this in a way that was discreet and casual."

These comments reflect what the students' experience of an ethics course might be. In other words, that it is important that there always be scope and safety in the classroom for personal choice and personal decision making.

The Purpose of "Teaching" Ethics

What is the purpose of activities such as discussing and debating ethical issues? Would it be reasonable to assume that if one works in an environment characterized by a "subculture" (discussed briefly in the introductory chapter and discussed in greater depth in Chapters 5 and 6), there may be considerable pressure on an individual to follow the norms of the group in which she works, and to do the accepted thing? If individuals choose not to yield to the subcultural pressures, the penalties can be severe. That being the case, would it again be reasonable to assume (and this is borne out by my research, discussed in the chapters mentioned above) that individuals from time to time may choose to act less ethically than they might do otherwise, simply because they consider the potential costs of doing what they think is the right thing to be too great?

What, then, are justice educators to do, particularly those who are concerned with the ethical aspects of working in corrections or policing? Should they simply inform you the student that this is the way things are, and that the subcultural constraints are a reality that has to be accommodated if you are to survive? Should they ignore this facet of work in this field and allow you to learn about the constraints the hard way? It is my experience that the more common approach is the latter one. Indeed, a quick, informal survey of police education and corrections worker students, and of police officers and corrections officers, indicated that the issue of subcultural constraints generally is not addressed in training programs. Some of these respondents, after my addressing the issue in ethics seminars they were attending, expressed concern in some cases, and anger in others, that this issue had not been addressed.

Often I have asked my own students what we should do. Would they prefer to be told about the justice field "warts and all," or would they prefer to receive a kind of "laundered" version, a version that is glamorous and all gussied up? Almost without exception the response is that we should tell it like it is. Not only should we do so, say the students, but to fail to do so would be improper and even unethical.

The risk for the trainer and educator, of course, is that you, the student or trainee, may develop a distorted view of what the field is like because you may put your own spin on the discussion, or you may lose some of your early enthusiasm for the officer's job, or you may approach the job more cautiously and defensively than is warranted.

The clear message I receive, however, is that those in the know about police and corrections officer roles should be upfront about the good and bad sides of the job, the advantages and the disadvantages, and that the risks of not being absolutely honest about all aspects of the job outweigh those of being honest.

Reflections

What is your take on this issue? Do you feel that your teachers should tell you about the potentially unpleasant and troubling aspects of working in the criminal justice field?

Should they attempt to heighten your awareness about group pressure and how it can impact on your sense of individual moral responsibility?

If you feel they should, is there any risk that this information may unfairly color your perception of the job, or lead you toward cynicism even before you get started in your career?

If you feel they should not, and that it is better for you to find out about the troubling parts of the job on your own, might you in some way blame them later on for not fully informing you?

Is There a Place for Ethics Education?

Let's return to the original question: Is there a place for ethics education in criminal justice training? As was stated earlier, quite a debate surrounds this issue, with equally convincing voices on both sides.

A report issued by the University of Alberta Task Force on Ethics (1985) and a subsequent progress review (1988) provide a useful framework for this discussion. They serve as a catalyst for examining some of the various perspectives on ethics education. The original report was based on the results of an extensive survey of opinions of faculty, students, and professional associations. The survey results led the university to commit itself to the idea that it should play an active role in ethics education (p. 67 following). Indeed, the first recommendation called for a required exposure to ethics education for all first-year undergraduate students:

> That the University require all first-year undergraduate students to take a course with an ethical component. This need not mean that every student must take one particular

course, but may involve a choice from a variety of courses which have been identified as having a significant ethical dimension. (p. 73)

Most important, the report makes it clear that its purpose "is not to promote the instilling of specific ethics, but to raise the level of awareness about ethical issues" (p. 24). This is a critical distinction to make, and it helps us develop a better understanding of the difference between teaching ethics *per se*, and providing an opportunity for learning about ethics and moral decision making. If I regard my role as one of "teaching ethics," then, as we said earlier, this implies I know best, in which case I would probably wish you to do what I suggest, or tell you, to do. It also raises the ugly possibility of what I call the lurking dangers of indoctrination. This topic is discussed in greater detail in Chapter 4 (see pp. 27 following).

The university survey results are interesting. Eighty-two percent of 1,025 students surveyed agreed that a "university education should help a student develop ethically" (p. 58). The professors' responses were recorded by department, the highest rate of agreement to this concept being found in the faculties of law (91 percent), and health sciences (89 percent). This is encouraging for those of us who feel that one day we may need the services of either a lawyer or a doctor! Or, might it indicate that these are the two professions most in need of ethical attention?

Opinions Are Split

As indicated earlier, and not surprisingly perhaps, there was quite a diversity of opinion expressed by the university professors in their written responses to the survey questions. More often than not, their comments were either completely *for* the teaching of ethics or completely *against* it. This is a topic that seems to attract strong opinions in one direction or the other.

The comments of those opposed to the notion of ethics education are particularly interesting. One respondent noted: "We must be careful about a university trying to shape ethics and morality for the community as a whole, because different people have different conceptions of values and ethics" (pp. 37–38). Another responded: "I don't believe it [the teaching of ethics] should be a goal in itself. The ethics systems of our university-age students are already in place and are basically unalterable at that age" (p. 38). One response was clearly conditioned by the fear of the possibility of indoctrination: "University professors are, other than in intelligence, like the general population, and include some people who lack moral judgement. In God's name, no potted required courses in ethics and values! This results in brainwashing, not critical thinking, in promoting 'acceptable attitudes,' not hard thinking" (p. 38). And finally: "No one becomes ethical merely by teaching or taking a course in ethics. Every course ought to be an ethical exercise" (p. 38).

The caution expressed in the first quote in the preceding paragraph is worth heeding. As we have indicated, we should take great care not to attempt to legislate a particular concept of morality, or to claim any monopoly on wisdom in this regard.

The second sentiment, adopting the view that students' values are pretty well fixed by university age, is a surprising one. Many people who are involved in any kind of teaching activity with adults speak about the way values can change as a result of the educational process. This would be particularly true of those students who may have simply "adopted" their value system from others, without thinking about those values in any real way. One of my students wrote in a journal: "Before coming into this [Ethics] class I felt that once you reached a certain stage your ethics were set, but because this course has made me look at myself, certain things have changed." And another wrote: "I have learned that my opinions and beliefs are not carved in stone, and may be changed if presented with a good argument."

The third sentiment again is one worth heeding. The legislating of ethics and values by an individual or organization necessarily forecloses on individual critical thinking. The danger here would be that of the individual being required to subjugate his own sense of morality in the interests of fitting in.

The fourth and final sentiment also requires serious consideration. To think of a course in ethics as some kind of panacea for all the ills of a particular justice system would be foolhardy. The view that every facet of a justice program should include consideration of ethical issues seems to be a more appropriate approach.

Nonetheless, most of the opinions in the report were in support of the idea of ethics education in principle, although there was some disagreement as to what form this should take.

Reflections

Are you surprised that 82 percent of the 1,025 students surveyed for the University of Alberta report agreed that the university should help them develop ethically?

What is your view on the opinion expressed by one professor that by the time students are of postsecondary age (generally 18 and up) their values are pretty well set?

One professor is clearly of the view that ethics courses are designed to promote " 'acceptable attitudes,' not hard thinking." What do you think of this point of view? (*Note:* This is a question that you could and should be asking yourself at this point in the course *and* at its conclusion.)

Two writers in the ethics area, Daniel Callaghan and Sissela Bok (1980), in their discussion of the doubtful goals of teaching ethics, affirm:

> It is often said that one test of the success of teaching ethics would be change in the moral *behavior* of students, and that a central goal should thus be an attempt to change behavior . . . this is an exceedingly dubious goal. First, even if a course could change behavior, it is hard to see how, short of constant reinforcement of the new behavior, its effects could be long or surely sustained once out of the classroom; other influences would play their role, and no course could be a permanent antidote against them. (p. 69)

In the criminal justice context, of course, one significant influence would be that of subcultural constraints. And these writers are in all likelihood correct in affirming that it would be highly improbable that an ethics course, in and of itself, could be a permanent antidote for those influences. My own experience of teaching ethics courses and conducting ethics seminars is that they do, indeed, have a "shelf life"; that is, that any effects begin to wear off after a while. Several students have talked to me about the importance of providing what they call "booster shots."

The discussion also raises the issue of the importance of thinking about supplementing ethics courses by encouraging teachers and students to consider the ethical implications of all elements of any educational program in justice departments. Consideration of the ethical dimensions of courses on, for example, Evidence, Arrest Techniques, Court Procedures, Security Techniques, or Use of Force Legislation, to name just a few, could build on learning and skills developed in a foundation ethics course.

Advice-Giving

Is there a place for advice-giving by your teacher in ethics courses? There are those who feel there may be a place for advice-giving as long as it is done carefully and sensitively by a credible and knowledgeable person. One of my teachers at the University of Toronto, in fact, advocated for such a viewpoint. Clive Beck (1991b), in *How Adults Learn Values,* wrote:

> People have a strong interest in the outcomes of their behavior and will adjust their values and actions in the light of information about consequences of behavior. Much of this information can be made available through instruction. (p. 14)

Michael Davis (1990), in an article entitled "Who Can Teach Workplace Ethics?" takes a similar view. One way of teaching ethics, he claims, is what he calls the "prudence" approach:

> You explain right and wrong in terms of what the employer, the law, or the profession wants and what will happen if one disobeys. You might, for example, explain why an employee should be prompt in this way: "If you don't want to get fired, arrive on time." (p. 30)

There may indeed be a place for this kind of prescriptive approach in criminal justice training, because someone who knows the expectations and fully articulates them can save a new officer from considerable grief later. It is particularly important to pass this information on to recruits, because only by so doing could they reasonably be held accountable for subsequent unacceptable behavior. It would be important here, of course, to inform students about policies and procedures, the mission statement if there is one, and the "lore" of the institution or the force—the unwritten rules of conduct governing corrections and police officers.

This kind of advice-giving about the rules and expectations of a particular workplace by teachers, however, as important as it is, is not the same as teachers telling you how you should respond to ethical dilemmas you encounter in the course of your job.

The Collaborative Process

Earlier we talked about the role of the teacher in facilitating the ethics educational experience. While there may be a place for teachers prescribing employer expectations in terms of behavior, it would be more consistent with a facilitative approach to learning for the teacher and student to engage in a collaborative process. This approach would be characterized by the teacher being both a teacher and a learner, and the student also being both a teacher and a learner. Clive Beck (1991a), in *Approaches to Values Education*, says:

> . . . a large proportion of the teaching is done by the learners, who instruct each other and their "teacher." The main point of values instruction, therefore, is not that one person tells other people what to value, but rather that there is explicit, systematic, joint inquiry into value matters. (p. 8)

In the business context, Dan Rice and Craig Dreilinger (1990) also favor the collaborative approach. The ethics experience, they affirm, should accomplish three goals: (1) heighten employees' ethical awareness; (2) provide them with the means to identify ethical issues; and (3) help them develop tools to clarify and resolve ethical issues (p. 106). There is no hint of "telling" or prescriptiveness in this view; the approach is one characterized by cooperation and helping on the part of the "teacher."

One of the reasons we should be careful about an overly prescriptive approach to ethics education is that as human beings we are each unique. Therefore, each of us necessarily brings to any moral dilemma subjective perspectives. Nel Noddings (1984), in her book *Caring: A Feminine Approach to Ethics and Moral Education,* talks about the "uniqueness of human encounters including those in the moral domain" (p. 5). She adds:

> Since so much depends on the subjective experience of those involved in ethical encounters, conditions are rarely "sufficiently similar for me to declare that you must do what I must do." (p. 5)

And Daniel Callaghan and Sissela Bok (1980), whom we mentioned earlier, in a book they edited called *Ethics Teaching in Higher Education,* also caution against a "telling" approach:

> No teacher of ethics can assume that he or she has such a solid grasp on the nature of morality as to pretend to know what finally counts as good moral conduct. . . . It is the time and place to teach them [students] intellectual independence, and to instill in them a spirit of critical inquiry. (p. 71)

Reflections

What are your views about having a separate "foundation" course in ethics, supported by you and your teachers, considering the ethical dimensions and implications of all your program courses?

Is there a place for advice-giving by your teacher in the context of an ethics course? If so, what kind of advice would you (1) expect to be given and (2) feel okay about being given? If you feel there is no place for advice-giving by the teacher, why not?

In the context of a criminal justice program, do you think there is a place for a collaborative approach to learning between you and your teacher?

Summary

We have discussed whether there is a place for ethics education in criminal justice training. We have suggested that because of police and corrections subcultural constraints on the autonomy of the individual, for these people taking time out for thought and reflection about ethical issues and moral dilemmas might be an important and wise use of time. We have suggested that stand-alone ethics courses lose much of their potential effectiveness if attention is not also paid to the ethical implications of all aspects of a justice training curriculum. We have suggested that the role of the teacher is an important one, but that how it is played out deserves some attention. We have suggested that there may be a place for advice-giving on the part of the knowledgeable and experienced teacher but that this should be engaged in cautiously. And finally, we have suggested that a collaborative approach to learning in an ethics course might be the less problematic approach in terms of negating the potential effects of teacher influence.

In the following chapter, we will consider this latter point in greater detail.

The Role of the Ethics Educator: The Lurking Dangers of Indoctrination

I sometimes found your mannerism was downright frustrating, because you did not help us make right choices. . . . You helped prepare us for possible situations we might encounter, and the dilemmas we might face in those encounters.

—Journal entry, justice student in ethics course

During the course a lot was said about teaching ethics and being influential. I don't think a person can teach ethics and not be influential.

—Journal entry, justice student in ethics course

I believe that a teacher has just as much of a right in stating his or her opinion and thoughts on any issue, as long as the students are aware that all people have this right.

—Journal entry, justice student in ethics course

I was sure that this course was meant to change my opinions and my beliefs. I also thought that you, the professor, thought your ethics were the correct ones, and planned to force them on us. However, something changed along the way and I began to realize there are two sides to every issue. Maybe my opinion wasn't always the right one.

—Journal entry, justice student in ethics course

I don't think that the teacher influences *what* I think, but I do think he influences *how* I think.

—Journal entry, justice student in ethics course

Chapter Objectives

After reading this chapter, you should be able to:

- *Identify and explain some dangers with respect to the role of the ethics educator.*
- *Describe the notion of* ***teacher influence.***
- *Explain why the teacher–student dynamic is a particularly awkward one in the context of ethics education.*
- *Identify at least one reason why students sometimes do not trust a teacher's invitation for them to think critically and independently.*
- *Identify four characteristics of critical thinkers.*

The Worry About Indoctrination

You may wonder why, in a book like this one that has been written for students like you, I have included a discussion on the role of the ethics educator. Let me explain.

Many ethics educators worry about the potential for their becoming doctrinaire in their approach to ethics education or for being perceived by their students to be so. This is a legitimate concern. That what I do in the ethics classroom will be perceived by my students as indoctrinatory has caused me some angst during the last 20 years since I started teaching ethics courses. The word *angst* is used advisedly. Much of the discussion that follows is based on my own personal experience and struggle. My hope is that this discussion and debate will provide interesting and provocative food for thought in your classroom. I would suggest that you think deeply about this important topic, and I would also encourage you to be as honest about it as you can be in any classroom discussion.

Clearly, my discomfort would seem to indicate that I see the words *indoctrinate* or *doctrinaire* as necessarily negative ones. I consider a doctrinaire approach one to be avoided studiously. Perhaps I should explain here, without getting into too much of a philosophical discussion, what I mean by *indoctrinate.* What I mean by the notion, first, is the attempt by one person (teacher in this case) to get another person or persons (students) to accept a particular position. Included in the notion, second, would be the intention by the teacher to withhold from a student the opportunity and right to consider that position for herself, and to develop her own perspective on it.

Instinctively I feel that there is something improper or inappropriate about one adult, especially one in a position of power as would be the case with a teacher, imposing on adults his views on various moral issues.

However, I have to confess that a significant problem for me is that I tend to hold strong personal views on some of the moral issues I discuss with my students. That is not to say I am inflexible; indeed, over the years I have come to accept different perspectives on many of the issues discussed because of the contribution of my students. On other issues, however, nothing that has been said by others has influenced my thinking. That being the case, it is a source of concern for me that some of my stronger views may in some way "leak out" during discussions with students.

Should an Ethics Educator "Come Clean" About Her Values?

It should be noted, of course, that some ethicists and educators feel that laying one's own views on the table is an appropriate and necessary thing to do. The knack, they claim, is to ensure that one's students are made to feel no pressure to accept those views and are encouraged to express alternative ones. Indeed, some are of the view that to refrain from doing so may be dishonest or manipulative.

A philosophy professor, during one of my postgraduate courses at university, used this technique of stating his views on each topic prior to class discussion. As a student, I was taken aback when he led off each discussion by arguing forcibly for his own view. I asked him whether he felt there was any danger of his foreclosing on dissenting student views because as the professor and the awarder of grades he might intimidate students. Did he feel students might feel obliged to acquiesce to his point of view? His response was that it was his responsibility to "come clean" about his own views, put them on the table, get them out in the open—but then work to create an atmosphere where students would feel comfortable in attacking those views if they could.

I am not entirely convinced. There! I have expressed a personal opinion on this issue. There is a distinct danger, in my view, whether one is a professor, or a staff trainer at a police or corrections training academy, of being perceived to be overly influential on the group. The teacher is invariably going to be seen by the students, and with some justification, to be in a position of power, and this is going to be an ever-present dynamic in the classroom. This potential for influence may be less of a problem at the postgraduate level, but it is still there; and at lower levels of education it is going to be very much there.

During one class with this particular professor I gathered up my courage and expressed surprise at his approach. Surely, it is a risky activity to argue his own views? Surely, it would be more appropriate to attempt to minimize, in whatever way he can, his potential for influencing his students? During the discussion, one classmate gently reminded me that, as laudible as it was that I was concerned about the dangers of too much influence over the thinking of others, it was inevitable. It was pointed out to me that the fact that I am male, white, speak with a British accent, and may be perceived by students as something of an expert, means I can operate from no position other than one of power. Try as a teacher might, therefore, it can be extremely difficult to create a climate of open, honest debate with his students, whether he comes clean about his views or not.

Sick Days

Perhaps at this point it would help the dialogue if we were to examine one of the ethical issues I discuss with my students. The issue is whether it is ethical for an employee to take a "sick day" when he is not sick. This is a particularly contentious issue that inspires some lively discussion in my classes. It is also an issue about which I hold some strong views. Simply put, I feel it is unethical for employees to take a sick day for any reason other than sickness. This is *my* value and it governs my behavior with respect to this issue. This is not to say that I am correct or that I have not and do not consider other perspectives. Frankly, however, my belief is that for me to take a sick day when I am not sick is tantamount to theft. And, other than for the obvious legal difference, I see little or no moral difference between taking a sick day when I am well and helping myself out of petty cash to whatever I am paid for a day's work.

Students tend to become defensive when we discuss this issue. They often engage in what I consider to be rationalizing. Sick days, they argue, are provided for in the employment contract and they are an employee's right. Who is to say what sickness is or what it is not? What if, despite my not being sick, I come back to work the following day and perform better as a result? All kinds of what I consider to be red herrings are drawn across the trail. Participants tend to struggle for a reasonable and defensible response when it is suggested to them that such provisions in the contract are there in case there is a genuine need to use them; that the sick day provision is made for *bona fide* cases of sickness.

Influence by Accident?

The challenge for me—and I constantly question myself about this—is whether given my strong feelings on this issue, and even though I do not express an opinion on the issue, it is possible for me to intellectually challenge others' ideas and perspectives in a dispassionate way? *It is this sort of question that is kernel to worries about indoctrination.* Not only am I a person holding power, but I am also a person holding strong views on certain important issues. We need to ask ourselves whether these are, potentially if not actually, two active ingredients in the indoctrination process.

What if, hypothetically, my views can be demonstrated to be morally sound? Does this give me the right to influence or attempt to influence the views of others; in other words, to have their thinking become more akin to mine? Is the role of the ethics educator one of deliberately setting out to make individuals more ethical; to have them think and act in a way prescribed by the educator? Or, is it more the mutual engaging in a process of critical thinking about, and examination of issues?

Fertile Ground for Influence

Let's say for the moment that the question of mutually engaging in a critical thinking process reflects more accurately what should be going on in an ethics classroom; that you the students are encouraged to think critically and to think for yourselves. Might there, however, still be a possibility of the ethics educator, especially one who is well liked by you, of seducing you into thinking in a given way? If the learning climate is a comfortable one, where the teacher is respected, trusted and well liked, might this be fertile ground for the overt or covert manipulation of your group to that teacher's own views? Let's consider another example.

Later in the book, reference is made to a conversation I had with a student when I was administrator of the justice department at the college where I work. The student was in the police science program and he had been caught cheating on an examination. When I asked him how he would square this activity of cheating with the trust and responsibility society would invest in him as a police officer, his reply was "It would be different if I were in uniform, I would be honest on the job." Several years later I am still trying to come to terms with this response and the kind of thinking and morality that would inspire it. The student clearly felt that it was okay to be dishonest in one's personal life, but not in one's professional life.

There appears to be little or no cognition of the possibility that the two may have some kind of connection.

This incident is often a topic of discussion in my classes. It is common for participants to defend this student's response. What we have here then is a situation where, in a real sense, I would like the participants to see that there may be something worrying about the student's response—in other words, to come around to my way of thinking. As a result of the debate and the concerns I raise, some may change their thinking and come to see that perhaps there *is* something worrying about the student's response.

What has happened here? Have I been guilty of indoctrination? Or, have I simply been an agent of change for those who have decided for themselves to embrace the change? Certainly, no one is prohibited from, or judged for, continuing to think in any way she chooses on the issue. The students are challenged intellectually, and are asked to offer reasons for their views. And even if they can't defend their views they are still, of course, free to continue holding them.

Comfort can be taken from the fact that wishing students would embrace a view I hold strongly may not be the same as inducing that embracing by way of psychological coercion or by overt or covert suppression of different ways of thinking. But the issues of power and control on the part of the teacher remain, and in my view need to be constantly guarded against. Sociologist Kimberley Folse (1991), in a journal article entitled "Ethics and the Profession: Graduate Student Training," refers to Churchill (1982), who makes the point that "moral values cannot fail to be taught" by teachers (p. 306). If this view has any merit, teachers find themselves in an unenviable position, through no fault of their own. It is as though they have had this (unwanted) mantle of responsibility placed on them.

It seems to me, from the many conversations I have had with my students on this topic that, like it or not, you are going to place this mantle of responsibility on us as teachers no matter how much we protest. My students tell me that, regardless of the amount of pleading I do that the last thing I desire is to influence them, this is exactly what is going to happen. First, they tell me I am their teacher, I am older, better educated—how could I not influence them?! Second, if you throw respect for the teacher into the mix, the potential for influence is magnified several times. Perhaps the ethics teacher who is going to be safest in this regard is the one who is not much older than the students, who is not well educated, and is not well liked! Of course, I am being a bit silly here, but you can see what I am getting at.

As much as I am discomforted by the thought that I am going to be influential, whether I like it or not my students make a very convincing case. It is important that you discuss this very important issue with your teacher or instructor. The following Reflections may help you in this respect.

Reflections

Should a teacher, particularly in an ethics course, be worried about being—or being perceived to be—doctrinaire by his students?

Would you prefer your ethics teacher to "come clean" about her personal views on a topic under discussion, or attempt to conceal them?

With respect to guarding against teacher influence, what do you think is the responsibility of the ethics teacher when one student says in class, "I think the trouble with this country is Chinamen and Pakis (sic)," or another says, "The native population is given too much money by this government, and they are all a bunch of drunken bums"?

Is there a place in the classroom for teacher influence if it is exercised in the challenging of racist or sexist remarks?

To what extent, if any, should teachers attempt to influence students who exhibit socially unacceptable attitudes?

If you think there is a place for a teacher to challenge socially unacceptable opinions, might this get in the way of the teacher being able to create an open climate where students feel they can state what they truly feel on a given topic?

What is your experience with respect to teachers and how they can influence you?

The Teacher–Student Dynamic in Ethics Education

As we have suggested, because of their role teachers necessarily find themselves in positions of power. They are in a unique position to influence. The potential for influence generally has its roots in either the student's fear of, or respect for, the teacher, or simply in the fact that how well the student does in the course is perceived by the student to be in the teacher's hands.

The teacher–student dynamic is a particularly awkward one in the context of ethics education because of the natural propensity of students to see teachers of any subject as experts. Often the relationship, somewhat sadly, is one characterized by the student seeking to "give back to the teacher what the teacher wants." Sometimes this may stem from the student's respect for the teacher's expertise, but more often it stems from the view that if one feeds back to the teacher what the teacher has given out, this is the best way to ensure a good grade. One of the more disappointing moments for me in working with students is for someone to approach me to find out what I "want" on an examination or in an essay. Such inquiries indicate that I have been unable to convey to the student that what I am really interested in is what *they* are thinking about a given topic.

It is a significant challenge, in my experience, to create a safe climate in a classroom where students feel they do not have to worry about agendas—including hidden ones—that the teacher may or may not have. In an effort to combat this fear, I commence any ethics seminar by informing the students that I make an honest attempt to harbour no agenda, hidden or otherwise; I attempt to convey to them that the classroom is a safe forum for the sharing of ideas, and that there are probably going to be no right or wrong answers regarding the issues raised. In other words, that the process of airing and exploring ideas is the important thing.

Traditional Education and Student Conditioning

The problem is that this kind of educational approach in which students are encouraged to think for themselves generally flies in the face of virtually all their previous educational experience. It is generally such an atypical educational approach that they can become suspicious of it, and some even intimidated by it. They have become so used to seeking and finding answers in the classroom that even if one asserts this is not part of the process here, they feel that this is what the teacher is about. Often they feel, probably because of the conditioning that has gone on in many classrooms during the previous 15 years or so, that perhaps in a subterranean or devious way, the intention is to attempt to lead them to think in a particular way. Often, attempts to disavow them of this notion fall on deaf ears or are met with a reluctance on their part to accept what the teacher is saying.

Leo Buscaglia, the American university teacher who has made a name for himself talking about "love," graphically depicts the problem with teacher-centered education in the foreword to his book *Love* (1978). "Education," he says, "should be the process of helping everyone discover his uniqueness. . . . But we still see the processes again and again of trying to make everyone like everybody else." He continues:

> A few years ago with some of my student teachers at the University, I went back into classrooms and was astounded to find the same things going on that had been going on when I was in school—a million years ago. For example, the art teacher would come in. Remember how we always anticipated and got ready for the art teacher? You put your papers down and you got your Crayolas out and you waited and finally in would walk this harried person. I really feel sorry for an itinerant art teacher. She comes racing in from another class and has time only to nod to the teacher, turn around, and say, "Boys and girls, today we are going to draw a tree." She goes to the blackboard, and she draws *her* tree which is a big green ball with a little brown base. Remember those lollipop trees? I never saw a tree that looked like that in my life, but she put it up there, and she says, "All right, boys and girls, draw." Everybody gets busy and draws.
>
> If you have any sense, even at that early age, you realize that what she really wanted was for you to draw *her* tree, because the closer you got to her tree, the better your grade. (pp. 20–21)

Buscaglia goes on to express concern about this tendency to demand conformity in the classroom, which, he says, continues "right on into the university." "We don't say to people, 'Fly, Think for yourselves.' We give them our old knowledge, and we say to them, 'Now this is what is essential. This is what is important.' "

If Buscaglia is correct, educators should not be surprised if students come into the classroom thinking that academic success depends on the degree to which they conform to the teacher's expectations. Presumably, this requirement to conform would apply to an even greater extent in police and corrections training academies because of the paramilitary nature of many of these facilities.

Is It Possible to Overcome Student Conditioning?

My experience is that there are ways to minimize these suspicions, but I have found that it is not possible to eradicate them. Of course, there are always students who, for whatever reason, seem to be more or less amenable to the new approach. Then there are those who, despite one's protests, continue to feel that the purpose of the exercise is to get them to think in prescribed ways; in other words, that the real purpose is to indoctrinate them.

For example, one of my students wrote:

> During the course a lot was said about teaching ethics and being influential. I don't think a person can teach ethics and not be influential.

And another student, in response to the different, more open, approach to learning, said:

> There is no way that a lifetime of conditioning [concerning previous classroom experiences] can be changed within a semester. . . .

Perhaps a more sobering comment is this one:

> You are well liked and held with great respect. I feel that people were afraid to say what they felt because they didn't want to lose your respect. In the back of our minds we knew this was wrong and you would never do it, but because of our conditioning through school, that was always in the back of our minds.

The latter comment is one that gives cause for concern for the educator, but it also provides, in a reverse way, a glimmer of optimism. The student who wrote the comment, (1) was able to express the concern, and (2) felt free enough to sign her name.

Here are some other student journal comments for you to think about:

> If the teacher did put his opinion into the conversation, I feel it would be a bad thing because some students would conform to his values and beliefs.

> There was never a time when the teacher was overly influential. At times he seemed to be concerned by the way certain individuals were thinking, but he never really judged us as unethical. Overall, I think the teacher was influential to the extent that that's what teachers are for—to be of some influence, to help us learn.

> If there was any time that the teacher was influential upon students, it was in their best interests. I think this is all right because he knows what we are going to be getting into and has already experienced it first-hand. He is only trying to prepare us to be successful.

> I was sure that this course was meant to change my opinions and my beliefs. I also thought that you, the professor, thought that your ethics were the correct ones, and that you planned to force your ethics on us. However, something changed along the way and I began to realize there are two sides to every issue.

Reflections

It would seem from these comments that despite my protests, I was only half successful in attempting to neutralize my influence over the students. What do you now think about this awkward and somewhat complex issue of teacher influence?

The Importance of Looking at All Sides

On the surface, what I seek to achieve in my interaction with students is to open their minds to alternative points of view on a given topic, where they are required to consider many sides of the issue before coming to a conclusion. Richard Paul (1992), who is a North American expert on critical thinking, in a chapter entitled "Ethics Without Indoctrination," argues that ethics can and ought to be taught, "but only in conjunction with critical thinking" (p. 240). He goes on to assert that in order to teach ethics morally, the course should be taught in such a way that students become "as adept in using critical thinking principles in the moral domain as we expect them to be in the scientific and social domains of learning" (p. 243). I agree. In my courses I spend time covering some of the principles and skills of critical thinking. I do so for two reasons: (1) I agree with Paul that it would be unethical to attempt to teach ethics to individuals when they are unlikely to have the capacity or permission to sort through the arguments themselves, and (2) it is a small but significant step toward guarding against the lurking dangers of indoctrination or the perception of indoctrination. The teaching of critical thinking, however, is more successful in protecting against actual indoctrination by the teacher than it is in protecting against the *perception* of indoctrination. Almost invariably, in my view, there will be students who, despite the teacher's best efforts, will become or will continue to be suspicious of the ethics educator's motives.

Here are a few brief comments made by three of my students about the importance of critical thinking. They are included for discussion purposes. The first one calls into question the idea that the educational process should be one that requires the teacher, as expert, to provide facts and information, and the student, in order to be successful, to simply regurgitate this material in examinations or papers:

> I have always had the notion that we never truly think while we are in college. All we do is memorize.

The second comment, even though it is expressed in the way that baseball legend Yogi Berra might have expressed it, contains an idea worth heeding:

> I did find the critical thinking part quite interesting, because it goes to show how much thought we don't put into our thinking.

And:

> It is easy to accept the *status quo* so as to not rock the boat. However, if throughout history no one thought critically about various policies and issues, and had not the courage to ask "why?," we might still be in the Dark Ages.

Indeed, had no one had the courage to think critically, perhaps we would still be convinced that the earth was flat, or that it was impossible for humans to fly.

Dialogue

Let's return to the indoctrination worry. Clive Beck (1991), one of my teachers referred to earlier, stresses the importance of dialogue in ethics education, and he sees this approach as one way of avoiding the possibility of the teacher being accused of being doctrinaire. He does caution, however, that: "Even with the dialogue format, authoritarianism may continue" (p. 3). He continues: "Careful planning and constant vigilance are needed to ensure that learners have freedom of speech, belief, and action and that the influence is reciprocal" (p. 3). Borrowing from Freire (1988) and Habermas (quoted in White, 1988), he lists the key elements of dialogue:

1. Respect for each other's insights
2. Respect for each other's "tradition" or "story"
3. Shared control of the form and content of dialogue
4. Focus on concrete, lived experience
5. Testing through action. (p. 2)

Noted sociologist Max Weber (1967), however, would probably argue that this vigilance against the dangers of indoctrination would not be a sufficient safeguard. He was writing in the context of science education, but one can only assume that his forebodings would be even greater in the ethics education context. He warns of the perils of teachers taking a stand on issues. For example, of political debate he says:

They [words] are not ploughshares to loosen the soil of contemplative thought; they are swords against the enemies: such words are weapons. It would be an outrage, however, to use words in this fashion in a lecture or the lecture room. . . . But the true teacher will beware of imposing from the platform any political position upon the student. (pp. 145–146)

He becomes almost strident in warning teachers off attempting to answer, or being drawn into, what he calls "questions of the value of culture" in their professional role. "The task of the teacher," he affirms, "is to serve the students with his knowledge and scientific experience, and not to imprint upon them his personal views" (p. 146).

Other Views on the Educator's Role

Two writers and researchers who have written extensively on ethics, Daniel Callaghan and Sissela Bok (1980), have some interesting things to say about this subject. Ethics courses, and by implication ethics teachers, they claim, "should make a change in the way students *think* about ethical issues . . . by providing them with the tools for a more articulate and consistent means of justifying their moral judgments" (p. 70). One of these tools would be the ability to think clearly, constructively and well. (A discussion of this particular tool and others is given in Chapter 6).

And Jonathon Baron (1990), in a journal article called "Thinking About Consequences," makes an interesting point about the importance of moral thinking:

Much of what we call immoral behavior, such as the behavior of Hitler's followers, or less insidiously, the followers of Colonel Oliver North, arose not so much out of malice but rather out of *thoughtlessness,* that is, *failure to think of consequences and relevant moral principles, or out of unreflective commitment to questionable ideologies.* (p. 77) [emphasis is mine]

Baron believes not only that there is a place for moral education, but that it is absolutely vital if we are going to develop students with the capacity to think for themselves about ethical issues and moral dilemmas encountered in their personal and professional lives.

One of the ethics teacher's responsibilities would therefore be to ensure that her students are encouraged to do their own critical thinking. It is important that students understand clearly the differences between critical and uncritical thinking. For example, I advise my students that critical thinkers

- tend not to jump to conclusions;
- examine the credibility of the evidence presented;
- are humble and fair-minded; and
- courageously face up to, and incorporate into their thinking alternative points of view if, after examination, these appear to be credible.

Uncritical thinkers, I tell them, tend to think and act in the opposite way.

Reflections

How would you describe your educational experience up until now? Have you, generally, been encouraged by your teachers to think for yourself?

What do you think about the differences between critical and uncritical thinkers? Where do you see yourself in this respect?

How do you respond to the views quoted from the journals of my students?

Would you agree, based on the student comments, that I was only half successful in my attempt to neutralize the potential for influencing my students?

What might be the implications of this partial failure on my part for your experience in an ethics course?

Do you consider yourself—generally, based on the brief references to critical thinking above—to be a critical or uncritical thinker?

Summary

How can we sum up this discussion about the role of the ethics educator?

We have discussed the worry about a doctrinaire approach to ethics education. We have discussed whether it is wise, or appropriate, for the ethics educator to "come clean" about his own views on a discussion topic. We have raised the question of whether it is possible for a teacher who chooses not to express her opinions to conduct a dialogue with students in a dispassionate way. Also raised was the dilemma for the teacher in establishing a safe, open climate for learning, while at the same time challenging socially unacceptable values. We also discussed the challenge that the ethics educator faces in creating a safe learning climate given to independent and critical thinking by his students, when most of the educational conditioning to date is characterized by the notions "teacher knows best" or "teacher as expert."

Subculture and the Individual Officer

So you either fit in, or you don't fit in. If you don't fit in you are not going to last. It's the same as the criminal subculture.

—Justice student after field practicum

One thing that really scared me was the need for acceptance within the group, which often leads to betraying your own sense of self, and the values we hold.

—Justice student after field practicum

The big dilemma was, do you stand alone and become a leper or do you let peer pressure and group think compromise your ethical stand?

—Justice student after field practicum

There is a price tag attached to honesty, especially in corrections. That is a scary thought. That even though you may be right, and that the actions of a fellow employee may have been wrong, in a subculture it is better to button up than to open up and speak to someone and risk being an outcast.

—Justice student after field practicum

I know I have spoken out in the past at work and I've suffered the consequences. Most times now I just keep quiet.

—Journal entry, professional person in ethics class

I can remember a specific topic from the class, "Does silence imply consent?" Maybe there is another question that should be considered. "Does silence imply fear?"

—Journal entry, justice student in ethics class

Chapter Objectives

After reading this chapter, you should be able to

- *Explain the significance of including a discussion of subculture in a book about ethics.*
- *Explain what we mean by the terms* ***subculture*** *and* ***subcultural constraints.***
- *List six norms that characterize police and correctional officer subculture.*
- *Identify two of the more significant norms in police and correctional officer subculture.*
- *Explain why police and corrections subcultural norms may be more compelling on officers than subcultural norms in most other workplaces.*

The Idea of Police and Correctional Officer Subculture

As you will have noticed, the first four chapters referred to the idea of police and correctional officer subculture. I have also alluded to the idea that subcultural constraints may inhibit individual officers from making their own independent, moral decisions. It is because of this latter sentiment that I believe any text addressing ethical issues in the criminal justice field would be incomplete without also addressing one of the more constraining forces against individual moral decision making—namely, that of a subculture.

In this chapter we will more closely examine the idea of a subculture as it pertains to police and correctional officers. What is a subculture? Is there a difference between a subculture in the criminal justice context and a subculture in other work contexts? What are its defining characteristics?

The discussion in this chapter and the next may touch a nerve. It has the potential to make you feel defensive. On the other hand, some of you may even feel a sense of relief that someone (stupid enough?) like me is prepared to broach the subject!

For you to keep an open mind about this topic, and for you to give it a fair hearing, will require considerable courage, and intellectual and emotional maturity. Of course, if after reading Chapters 5 and 6, and thinking about and discussing their contents, you wish to discount some or all of the ideas, then that is, of course, your right. What matters most is THAT you think. But be aware, it will not be easy for you to keep an open mind. I suspect, based on discussions of this topic with students, that it may be much easier for you to react defensively.

What Do We Mean by "Subculture" and "Subcultural Constraints"?

To help us develop a better understanding of the terms *subculture* and *subcultural constraints,* we consider three sources of information. First, we turn to a dictionary definition for help. Second, we discuss the ideas that some faculty members in college justice programs have about these two terms. Third, we look briefly at what some researchers have to say about them. In the following chapter, we will consider the views of some justice educators, police and corrections officers, and administrators.

Definition of Subculture

First, let's look at a dictionary definition of subculture. The *Collins English Dictionary* (Third Edition) defines subculture as "a subdivision of a national culture or an enclave within it with a distinct integrated network of behaviour, beliefs, and attitudes." The key idea here is that a subculture is characterized by a code of behavior governed by a set of beliefs and attitudes.

Faculty Opinions

Second, let's consider what faculty in college criminal justice programs have to offer in helping us understand these terms. It is significant, I think, that all of these faculty members have enjoyed careers in the justice field, and since becoming educators have supervised students extensively on their field practica. This activity has enabled these teachers to keep up with developments in the field. It would be reasonable to assume, therefore, that their opinions continue to be current and informed.

The faculty opinions were expressed in a survey that I conducted. Faculty were asked to respond to the question "What do you mean by the term *subculture?*" Their responses, I believe, were interesting and informative and will provide stimulating topics of discussion for you and your class colleagues. Keep in mind that in all cases the faculty were defining subculture as it relates to the justice field. (In the next chapter we will consider what these justice educators said to me about this topic in one-on-one interviews.)

As you read this chapter, take some time to reflect on your own personal experience. What you think and feel about your workplace and how it impacts on you as an individual is of critical importance to your development as a human being and as a police or corrections officer.

Here are some samples of survey responses. One teacher wrote: "By subculture, I mean a different, distinct, and separate cultural phenomenon—a culture other than the mainstream culture. One with its own values, beliefs, mores. The major value of this subculture is to 'protect your own butt', and, secondarily, 'to protect and stand by your coworkers.' "

It is interesting to note what this person claims are the two main values characterizing police and corrections officer subculture. Clearly, these two values more often than not characterize *any* workplace. The difference, however, between, say, a criminal justice workplace and an automotive plant or an insurance office, may be the *degree* to which these norms characterize those particular workplaces. As we shall see later in this chapter, arguments can be made that because of issues around the safety and security of officers in the justice field, the requirement to adhere to subcultural norms may be stronger.

"The subculture," claims another faculty member, "consists of a set of beliefs, habits, attitudes, and even schemes which determine in large part what line workers [officers] have to do to survive in their occupational environment." Interestingly, this person added in his response that in such a subculture these beliefs and attitudes are conveyed *informally*. They constitute essentially what is a code of conduct for officers, known by all officers but normally not formally communicated. The code becomes understood, almost by osmosis, and then is reinforced when a nonconforming officer is chastised (ostracized?) by her colleagues. We referred in Chapter 1 to the view of Professor Frederick Desroches that "novices [new police officers] entering the group are confronted . . . with a set of norms and values to which they are exposed and are expected to conform." There are some clear connections here between this particular faculty member's ideas and those of Desroches. The faculty member talks in terms of what "officers have to do," and Desroches talks about an expectation of new officers to conform to what we sometimes call a code of conduct. Both these people make it sound as though one doesn't have much choice in the matter. Perhaps, as we noted in Chapter 1, that is why two researchers, Grossi and Berg (1991), talked in terms of corrections officers having to "compromise their personal values and interests." If this kind of thinking has any merit, isn't it curious that an informal code can be so powerful?

We have referred to an unwritten, informal code of conduct governed by a set of expectations. One faculty member, in explaining his views on subculture, talked in terms of "a group of persons who adhere to an unwritten code of conduct which governs certain areas of behavior. The subculture is also the vehicle through which values and morals are conveyed by their proponents, and the rewards/punishments imposed for conformity/nonconformity."

Many other faculty members responded in a similar vein to the survey question asking them to define what they mean by *subculture.* A recurring theme was that a subculture is generally characterized by expectations that the group has regarding the behavior of its members.

Researcher Opinions

In a recent ethics training class, participants were discussing the manner in which the police "subculture" impacts young officers, when the chief of a small police agency in the state of Illinois made a very interesting observation. He compared the indoctrination process for new officers seeking entry to the police subculture, to youngsters seeking membership in the street gang subculture. "There is really only one essential difference," the chief said. "The street gang tells those seeking membership exactly what is required of them. The police subculture does not."

Many of the researchers in this area tend to agree with this kind of thinking. Sociologist Theodore Mills (1967), for example, talked about the norms of the group being "ideas in the minds of members (of that group) about what should or should not be done by a specific member under specified circumstances" (p. 74). Three other researchers—Alan Coffey, Edward Eldefonso, and Walter Hartinger (1982)—in discussing police subculture, defined it "as a kind of shorthand term for the organized sum of police perspectives, relevant to the police role" (p. 141). Two others, Darwin Cartwright and Alvin Zander (1968), talk about the pressure that groups can bring to bear on their members: "But it is clear that groups can, and often do, apply pressure on their members, so as to bring about a uniformity of beliefs, attitudes, values, and behavior. . . . If a cohesive group has developed a standard or a norm, it may exert strong pressures on any member who attempts to deviate" (p. 145).

Perhaps you will agree, if any of the foregoing ideas are worthy of serious consideration, that a discussion of subcultural constraints has a place in a book about ethics. If the group to which we belong, in the interests of maintaining group solidarity, expects us, even occasionally, to give up our right to make our own moral decisions, maybe we should be concerned.

Of course, many other aspects of the development of a subculture need to be taken into consideration, and again the researchers can help us develop an understanding of these.

A Subculture Needs Perceived Enemies

Some researchers express the view that for a subculture to become established, the subcultural group has to be anti-something or another. In other words, that there has to be a common "enemy." Members of the group join together in their opposition to the "enemy" and this, in turn, can produce the effect of galvanizing the group. For example, sociologist G.L. Webb (1978), in addressing corrections officer subculture, talks about that particular group being anti-inmate, anti-administration, and anti-prison social worker. Criminologists Elizabeth Grossi and Bruce Berg (1991) talk about "hostility directed at officers by inmates, their [inmate] families, and often even the public," thereby making them the "enemy."

In the police context, similar sentiments have been expressed by other researchers, including Bittner (1980) and Goldstein (1990). Police officers, for example, when they perceive themselves to be under siege from the public over an alleged act of professional misconduct, often see themselves as being misunderstood. In other words, they feel only a police officer can understand police conduct—"they" (the public) do not understand "us." In the following chapter there is a discussion of one former police officer's description of what he calls the "musk ox syndrome." His depiction graphically portrays how members of a police force may react in the face of public scrutiny or criticism. It doesn't constitute a quantum leap, I think, to see how this kind of dynamic can have the effect of galvanizing a group.

In summary, then, we have briefly examined what we mean by a subculture in the justice field. We have suggested that it is characterized by a set of norms, communicated informally *to* all of its members *by* members. These norms are said to govern the conduct of those members if they wish to preserve their membership in the group. We have also discussed the idea of a subcultural group needing—for its formation and continuing existence—to have a real or perceived enemy. We have suggested that this can galvanize the members of the group into a "solid" entity. In the case of police and correctional officers, the "enemy" is often seen to be the criminal element, the officers' respective administrative groups, and, perhaps to a lesser degree, the public.

Reflections

Would you agree that the issue of police and corrections officer subculture is a highly sensitive and emotive one? If so, why do you think this is so? What feelings have you been having since you began thinking about the topic?

How would you define *subculture* in the criminal justice context to someone who had no idea what you were talking about?

What are your views on the idea of an unwritten code of conduct governing the behavior of police and corrections officers at work?

Are Subcultural Constraints the Same in every Workplace?

You will have noticed, for the purposes of this book, that the notions of *police officer subculture* and *corrections officer subculture* are used interchangeably as though they are, to a large extent, one and the same thing.

More research has been done on the former so we know a little more about subculture in the police context. From this research it may be argued that the subculture in policing is stronger than it is in correctional institutions, and that the influence of the subculture tends to remain relatively constant across differing police authorities. The idea of a police brotherhood is believed by most to extend across police departments. Perhaps that is why in countries like Canada, the United States, Australia, and Britain, to name a few, it is customary in sad instances of a police officer fatality for officers from other forces to attend the funeral to express support and solidarity. It is also common for correctional officers to send delegates to such funerals.

Subculture Is Strongest in Maximum Security Institutions

With respect to corrections, my research indicates that the strength of the subculture tends to correlate with the security level of a correctional facility. It tends to be strongest in maximum security institutions, and becomes weaker as the security level is reduced. However, even in minimum security institutions, an officer subculture is still evident, although the constraints are generally not as strictly enforced. For example, it is conceivable that officers working in minimum security facilities may see the inmate population not so much as the "enemy" but as a group of individuals who may benefit from their incarceration. Generally, this is most unlikely to be the case in maximum security institutions, and somewhat less likely to be the case in medium security facilities. This view was borne out in a discussion I had with an institution administrator. He said: "The nature of the subculture and how ingrained it is, in my opinion, depends on whether you are in maximum security, medium security, or minimum security. It is just like the inmate subculture which is highly ingrained in maximum, less so in medium, and less in minimum."

That having been said, police and corrections officer subcultures share, in my view, a number of common characteristics. These allow us to regard them as sufficiently similar to warrant considering them together. For the sake of simplification, then, and to avoid the risk of becoming bogged down in a debate over the finer differences between police and corrections officer subculture, or over the differences between subculture in a maximum security setting and a less secure institution setting, we will consider them to be essentially similar.

There may be a place in another book for a discussion about the differences, but here we are simply looking at the idea of police and corrections officer subculture in general terms, and how its constraints may influence the individual officer's sense of autonomy in the workplace.

Distinguishing Characteristics of Officer Subculture

What, then, are some of the distinguishing characteristics of police and corrections officer subculture, and why is it sometimes claimed that this subculture is different from that in most other workplaces?

Support for seeing the two subcultures as similar is offered by Elizabeth Grossi and Bruce Berg (1991), to whom we referred earlier. They write:

> Those who have studied the internal workings of correctional institutions may recognize that powerful forces within the correctional system have a stronger influence over the behavior of correctional officers than the administrators of the institution, legislative decrees, or agency policies. These social forces may be understood as related to a "correctional officer subculture," similar to that found in policing. (p. 79)

These two writers list what they call *setting-related stressors* for corrections officers that they claim are "remarkably similar to those commonly associated with police officers" (p. 79):

1. The ever-present potential for physical danger.
2. Hostility directed at officers by inmates, their [inmate] families and often even the public.
3. Unreasonable demands and expectations on their role as corrections officers; vacillating political attitudes toward the institutional role of corrections.

4. A tedious and unrewarding work environment.
5. The dependence corrections officers place on one another to effectively work in the institution safely.
6. The reality that one cannot always expect to act the way either one would choose to, or the way the public might expect them to.

(Point 6 is one of the few references I have found to support the idea that police and correctional officer subculture may impair the ability of the individual officer to make independent moral and practical decisions at work.)

Kelsey Kauffman (1988) is a scholar who researches primarily in the corrections context. In her book entitled *Prison Officers and Their World,* based on a four-year study of the Massachusetts prison system, she says, "Officers also possess a distinct subculture within prisons. Their own beliefs and code of conduct set them apart from administrators, social workers, and, of course, inmates." She "discovered" the beliefs and the norms that constitute the code of conduct by asking officers two questions: (1) What is the worst thing you could do in your own eyes as an officer?, and (2) What is the worst thing you could do in the eyes of other officers? Nine norms were identified:

1. Always go to the aid of an officer in distress.
2. Do not "lug" drugs.
3. Do not "rat."
4. Never make another officer look bad in front of inmates.
5. Always support an officer in a dispute with an inmate.
6. Always support officer sanctions against inmates.
7. Do not be a "white hat." (This denotes any behavior suggesting sympathy for or identification with inmates.)
8. Maintain officer solidarity against all outside groups.
9. Show positive concern for other officers. (p. 85)

Would it be reasonable to suggest that if we substituted the word *criminal, offender,* or *suspect* for *inmate,* many of these norms would apply equally to policing? Clearly, the predominant theme here is that of *solidarity* of the officers, and of support and protection of one another in the face of the common "enemy." The enemy is the criminal element and, to a lesser extent, outside groups and the public.

Let me offer an illustration of this point about officer solidarity. A newspaper article entitled "Four Officers Suspended in Probe" indicated that four police officers had been suspended pending a probe into allegations that they had planted drugs on suspects (*The Toronto Star,* March 17, 1995, p. A6). The article included the comment: "Detectives from the force's internal affairs unit launched their probe two weeks ago after a constable—whose identity is a closely guarded secret—complained that a veteran officer from 12 Division's B Platoon had been involved in illegal conduct. . . . Meanwhile, the name and division of the police officer who made the complaint are being kept secret. Police sources say they don't know what motivated the constable to come forward with the allegations at this time."

It would be inappropriate to read too much into this newspaper account. However, on the face of it we may have an example of one police officer feeling the need to "break ranks" and report behavior of colleagues that he considered unacceptable or even illegal. That the reporting officer's identity was kept a "closely guarded secret" should come as little or no surprise because of the potential for being ostracized by the group. (As an aside, there is a conundrum here for the administrator to whom the officer reported the wrongdoing. The administrator has to weigh keeping the confidence of the reporting officer against the right of his or her accused colleagues to know who the accuser is. We shall return to difficulties like this one later when we discuss the ethical dilemmas.) If Kelsey Kauffman is correct, this reporting officer failed to comply with at least eight of the nine subcultural norms. Please look at the norms and identify for yourself how many you think were breached by the officer. He is guilty of "breaking ranks," of not remaining "solid" with the officer group.

In August 1997, an outrageous example of police misconduct occurred in New York City. In that case, several officers were accused of participating in torture, including the insertion of a broomstick-like device in the rectum of a Haitian immigrant they had arrested. Shortly after this bizarre incident became public, it was reported that two officers had stepped forward, indicating their willingness to testify against those who had committed this horrific act. Despite the inhuman nature of the deeds with which several officers had been charged, it was interesting to note the manner in which the media reported about the two who "stepped forward." Almost every time their names were mentioned, it was pointed out that "they are under police guard for their protection."

You are correct if you are thinking we haven't yet identified what makes a subculture in the justice context substantively different from that in other workplaces. One writer, Vernon Fox (1983), talks about one difference in his book *Correctional Institutions.* He notes that much has been written about the loyalty of police to their colleagues and affirms that such loyalties are also common in the corporate world. "Nevertheless," he says, "this loyalty appears to be stronger and have more meaning in the criminal justice system than in most other groups" (p. 162). Some of you might disagree with Fox. You may argue that the subculture in General Motors, for example, is similar in many respects. You may well be correct.

One important difference, however, is the anti-criminal element. Most other workplaces see the customers in positive terms, often as "friends." If they did not, they would probably not survive for very long! Offenders, on the other hand—the "clients" of police and correctional officers—excite a different response, normally one characterized by antipathy. And this antipathy can create a bond among police and correctional officers as they see themselves working together against the "bad guys."

One other even more significant factor may account for the difference between subculture in corrections and policing, and that in other workplaces. In policing and corrections there is an ever-present potential for danger—life-threatening danger. The potential for danger exists because of who the "clients" are. Significant issues of personal safety are present that generally do not characterize most other workplaces. And it is this ingredient of danger that may lend credibility to the view that police and correctional officer subculture is stronger.

If the subculture is stronger, it may be reasonable to think of it as a more compelling constraint on the conduct of its members.

The Presence of Danger Reinforces the Subculture

Some interesting work was done in the 1960s on group solidarity by sociologist Irving Janis. Janis, basing much of his theorizing on war incidents, supported the idea that the presence of danger tends to increase the solidarity of a group. For example, he said: "It has long been known that when people are exposed to external danger they show a remarkable increase in group solidarity. That is, they manifest increased motivation to retain affiliation with a face-to-face group and to avoid actions that deviate from its norms" (1968, p. 80).

This, I believe, is the key difference between police and corrections officer subculture, and subculture in most other workplaces. The individual officer's welfare, if not her life, can sometimes rest in the hands of a partner. The willingness of colleagues to respond, support, and provide "backup" in an emergency is a basic tenet of the police and corrections officer's role. This, I would suggest, can in turn lead to a bonding that goes beyond the kind one might expect in most other workplaces.

Ratting: The Loyalty Norm

Security issues and the way in which individual officers have to rely on each other for their safety makes the requirement to remain loyal to one another the key norm in their subculture. There is a strong prohibition against what is often called "ratting" or "ratting out" of a colleague. Coincidental with writing this section of the book, I was conducting a seminar at a secure, young-offender facility. The officers and I engaged in a conversation about this topic. At the end of the day one of the staff came up to me, and surreptitiously pulled from his pocket his electronic key fob, on one side of which he had placed a picture of a rat. He had reported a colleague for hitting a resident, had been labeled a rat, and had decorated his key fob in this way as a humorous attempt to neutralize the stress of his "label." It is not unknown in corrections or policing for rather severe consequences to be inflicted on a colleague who has been labeled a "rat" by his colleagues. These consequences can range from being "put on the grease" (shunned by colleagues) to threats of personal or property harm being made—or carried out—against the "rat."

The television news program *60 Minutes* carried a story that may be pertinent to this discussion. According to the story (CBS News, March 30, 1997), two officers named Rigg and Caruso at California's Corcoran State Penitentiary blew the whistle on what they considered to be unethical conduct by their colleagues. The two officers claimed that inmates from different ethnic groups were set up to fight gladiator-style in the small exercise yard of the Special Handling Unit. It was not unusual, they claimed, for officers to place bets on the outcome of the fights. When the fights got out of hand, they claimed, officers would on occasion shoot at the inmates. Since the institution had opened in 1988, eight inmates had been shot dead by officers and numerous others had been wounded. Inmate Preston Tate, it was claimed, was told after he had hit an officer that he would never get out of the facility alive. On the program, videotape footage was shown of Tate and another inmate of the same ethnic origin fighting two other inmates of a different origin. Tate was shot and killed in the melee even though the videotape clearly shows that he was not the aggressor and that he was away from the fight when he was shot. Lieutenant Rigg, one of the whistle-blowing officers, described the shooting of the inmate as murder. In 2000, the investigation into this institution continues.

Whether the allegations are true has not yet been determined and is immaterial to the point I am about to make. During the interview with the two officers, both said they had been labeled as "rats," as "no-goods," and both had had their lives threatened by officer colleagues. Even though they had been transferred to other institutions the labels traveled with them, which made their lives in their new workplaces as miserable as they had been at Corcoran. At the conclusion of the story, interviewer Mike Wallace, at the new place of work of officer Richard Caruso, hailed down the vehicle doing the institution perimeter check and asked the officer driving the vehicle if he would shake hands with Caruso. It was a telling moment when the driver looked incredulously at Wallace, shook his head in a gesture of defiance, and drove off. Breaking ranks and being labeled a rat can be an unpleasant, unsettling experience.

In the mid-1980s, a female corrections officer at a medium security correctional institution reported some of her colleagues for sleeping on night shift. She had reportedly approached them and said several times that she felt unsafe and vulnerable when they were asleep. She told them that if they did not refrain from doing so, she would have no alternative but to report them to the superintendent. They continued. She reported them. The consequences for her were severe. Graffiti was written about her on some of the walls in the institution; she received harassing phone calls and letters; her car was deliberately damaged in the institution parking lot. The campaign of terror by her colleagues culminated in some bricks thrown through the windows of her home. It appears she paid a heavy price for doing what she considered to be the right thing. In this case, the right thing for her was to attempt to ensure her own safety and that of the institution.

It was many years ago now, but I remember experiencing a three-week ride-along with a police force. In a ride-along program, as many of you know, a civilian is taken along in a cruiser to observe police officers at work. I had been invited to do so after spending several weeks conducting in-service training sessions for this particular police force. I had also got to know several officers through a criminology course I was teaching at a local community college. The officers had accepted me and became very open with me about their work, about their rites of passage and about what they considered the "perks" of their job. I remember riding along with one particular officer who appeared not to be well liked or well respected by his colleagues. I noticed that the other officers were invariably slower to respond to his calls for assistance. I questioned them about this apparent antipathy because I could not make any sense of their reaction to this officer. He struck me as being a professional, principled, family-oriented individual. In retrospect I can see that he was deemed by his colleagues not to fit in. Several years later, I think that I can see why they may have reacted to him in this way. He was not one of "them." He marched to his own drummer. He refused to participate in the on-duty activities of some his colleagues, some of which, I remember, stretched the limits of professional conduct.

If we were to select those who should appear in the pantheon of officers who have placed themselves at risk for reporting the misdeeds of others, Frank Serpico would be high on the list. As a member of the New York City Police Department, he placed himself in great peril for exposing the rampant corruption within that agency in the early 1970s and for providing exact details about that corruption before the Knapp Commission. Today, more than a quarter century after that testimony, Frank Serpico continues to excite considerable emotion in large parts of the police community. Most officers readily describe him as a hero, while many continue to see him as a "rat."

This will probably not come as any great surprise to you, but I think it is interesting that the inmate subculture is also characterized by what we will call the *loyalty norm.* Anyone who has worked in the justice field, particularly those of you who have worked in correctional institutions, will be familiar with the way in which inmates are quick and sometimes merciless in weeding out from their number those they consider to be "rats." This is one of the main reasons most institutions have protective custody or super-protective custody wings, ranges, or cell blocks—so that inmate "rats" can be segregated for their own protection.

Perhaps you will permit me to make a personal observation. In my ethics seminars I discuss the issue of subculture and the police or corrections officer participants are generally quick to identify the "ratting" norm as a key one. In addition to the term "rat," they use, almost interchangeably, terms like "narc" or "fink." In Australia, the act of reporting a colleague is often called "dobbing in a mate"—a sin in the Australian culture considered almost equivalent to blasphemy! (It would be interesting for someone to investigate whether or not this strong societal norm resides in Australia's history of being settled by convicts.) In the seminars, the participants, when asked about the effects of the "ratting" norm on their individual sense of what may or may not be the right thing to do, become deeply reflective. And this happens almost without exception. It is as though the participants are taking personal stock of where they are as individuals with respect to this issue. When I question them about the apparent change of "chemistry" in the room, they generally confirm that this is precisely what is happening—that they are engaging in a personal check regarding how much of their individuality they may have conceded to their colleague group.

A Need for a Sense of Perspective

It would be wrong to discuss this issue of officer subculture without pointing out that the subculture has some extremely positive qualities. The mutual support and protection it is likely to engender in crisis situations is essential to the well-being of the officers. The camaraderie often experienced following a crisis situation is sometimes essential to the emotional and psychological health of officers involved. The feeling of belonging to a group whose members understand the peculiar and difficult demands of the profession can be a gratifying and comforting experience, especially when the group or the profession feels under siege from the public.

This camaraderie and support that police officers have with and for each other is graphically expressed in an article entitled "Courtroom 'Tense' for Detective in Shooting" (*The Toronto Star,* March 7, 1997, p. A6). A detective, charged after a teen was shot in the head and killed, was supported by "more than 100 off-duty York Region officers."

One criminal justice educator I was speaking with talked about this positive side. "The subculture at times can be a strength. At times, you may be asked as an officer to do some fairly disgusting things. . . . Once the incident is over, you still have your 'family' to support you." Continuing this thought, another officer said, "The subculture serves you well. There is a lot of camaraderie. I think people do look after each other. There is a recognition coming from one's peers that is more valued than that coming from administrators."

It would be inappropriate to expect individuals involved in the dangerous professions of policing and corrections to "go it alone." Indeed it would be, and is, a distinct worry for both colleagues and managers to have to work with a maverick, "loner" officer.

This chapter and the next are not about breaking the subculture. They simply raise the question as to whether, on occasion, the officer subculture can have a darker side; a side where too much of a person's sense of independence and individuality is given up to the group, even when the group may be acting against the moral suasion of that individual. Remember, in Chapter 1 we called it *de-individuation.*

Summary

To summarize, in this chapter we have discussed whether there are sufficient similarities between subcultures in policing and corrections to allow us to consider them, for the purposes of this book, as essentially similar. It was suggested that an argument could be made for doing so, although the finer differences between them would make for an interesting research topic.

By way of explaining the similarities between the two subcultures, certain norms, probably common to both, were identified.

The essential similarities are created and reinforced, we suggested, first, by the inherent danger in both roles—there are distinct and ever-present risks to the safety and well-being of the officers; second, by the reality that in these two subcultures the general feeling toward the "client" is one of antipathy, a situation not generally found in other workplaces in government, industry, or commerce; third, by a shared antipathy toward the public "them" who do not understand "us."

Because of these factors we have suggested that policing and corrections subcultures are, in some significant ways, different from "regular" workplace subcultures, and that the most significant difference—the threat to the officers' safety—ensures to a greater degree than in other workplaces that the loyalty norm is respected by individuals and enforced by the group.

Reflections

What are your views about whether police and corrections officer subcultures are generally similar?

What are your views about the idea that there are some significant differences between police and corrections officer subculture, and subcultures in most other workplaces?

What do you think of the nine corrections officer subcultural norms identified by Kelsey Kauffman and the six stressors identified by Elizabeth Grossi and Bruce Berg? Which ones, if any, do you think would apply to police officers?

Subculture: What the Practitioners Think

I know what the officer code is and it is very strong. A person can come into this work with all good intentions, but a lot of people just give up and fall into the pit [the subculture].

—Senior federal institution administrator

I would say the peer pressure is very strong. Very few officers would go against that peer pressure.

—Senior institution clinician

[If you are not loyal] you would probably be driven out of the institution very quickly. You would have so much pressure on you that you would resign.

—Corrections officer

It's important to be a member of the brotherhood, whether you're male or not, it's a brotherhood, you subscribe to the rules of the brotherhood . . . and you do not violate the rules. Time and again we see people being ostracized because they have broken the rules of the group. You had better not break the code of silence.

—Police trainer

I was once told by an officer I respect, "We have to be our own person and develop our own methods of work, yet we must manage these ideals within the framework of the institution."

—Criminal Justice student after field practicum

Chapter Objectives

After reading this chapter, you should be able to

- *Explain why it may be risky for police or corrections officers to talk openly about the group norms that govern their conduct.*
- *Identify three key ideas that practitioners have about officer subculture and subcultural constraints.*
- *Explain why subcultural constraints may make it difficult for officers to make autonomous moral decisions at work.*
- *Explain the terms **loyalty** and **solidarity** in the context of a criminal justice subculture.*
- *Explain why there may be a link between officer and offender subculture.*
- *Identify one reason why some officers may be able to take moral stands on certain issues and not be ostracized by their colleagues.*
- *Explain why members of a subculture need to identify common "enemies," and what effect this has on the group.*

A Sensitive Topic

Much of the material for this chapter has been gathered in one-on-one personal interviews. The interviews were conducted with criminal justice educators, police and corrections officers, and correctional institution and police administrators. The transcripts of many of the interviews constitute a part of the qualitative data gathered for my doctoral dissertation. The material is included here with the written consent of the respondents. Great care has been taken not to identify any police force, correctional jurisdiction or institution, or individual officer. The willingness of these people to speak to me is much appreciated. They have contributed significantly to my learning, and I hope their insights will contribute to your learning also.

The issue of subculture in policing and corrections is a highly sensitive and emotive one. In their willingness to express their views on this topic, some of these individuals, particularly officers, took extraordinary risks. The nature of these risks may become more apparent as the chapter unfolds.

To illustrate the risk taking, let me recount part of a conversation I had with one officer. As we were concluding the interview I asked the officer if he had any last thoughts about our discussion. The officer confided: "It is difficult. I have found this a difficult interview. I'm finding it (the subculture) is something that you don't talk with everybody about, and it is just, I don't know it's. . . ." "It's a question of trust," I finished, "and you don't know me from a hole in the ground." "That's it right there," the officer replied, "I think, too, because I am at work I have this shield up around me to protect me. . . . The coworkers are the only ones you can really trust."

Another officer, in response to my invitation to make some final comments on our discussion, said: "This had better not get out or I'm in deep doo-doo."

It should be pointed out that the officers interviewed were selected randomly; it was not the case, by way of allaying any suspicion you may have about the process, that these officers were handpicked in some way. When you read some of their comments, you may be tempted to think they were selected for interview because they had been identified as loose cannons or had already been targets for ostracism by their colleagues. This was not the case, either. Before interviewing them I had little or no idea of what their perspectives were going to be, or what risks they would take in sharing their ideas with me.

This chapter is organized around the various points of discussion in the preceding chapter. We continue to explore the terms *subculture* and *subcultural constraints.* We discuss in greater depth whether police and corrections officer subculture is different from subculture in other workplaces. We continue to explore the links between officer subculture and criminal subculture. We also look at the idea that the subculture is stronger in the criminal justice context and that, as a result, the norms are more compelling on individual officers.

It should, of course, be kept in mind that this discussion does not constitute the definitive word on any point discussed. It is included here simply as a means of fleshing out the discussion in Chapter 5. By including and thinking about some of the ideas of police and corrections practitioners, we may be able to develop a broader understanding of the issue. In this chapter, their perspectives have been interlaced with researcher ideas and with some case studies, by way of providing a context for these perspectives. Hopefully, the discussion will serve the purpose of providing further background for your thinking and debating.

"Subculture" and "Subcultural Constraints"

Throughout the interviews with the educator, officer and administrator groups, certain terms kept recurring:

bonding	adherence
code of silence	code of behavior

officer code	what officers have to do
unwritten code	conformity
sticking together	nonconformity
loss of individuality	loyalty
put on the grease	ratting
family	social control
ostracized	peer pressure

These terms are referred to consistently. They are, I would suggest, key to our understanding of subculture in the justice context. We should acknowledge, of course, that they can also be key to subcultures in other workplaces. What we have to keep in mind is the *degree* to which these terms characterize a particular subculture. As I have suggested, and this is borne out by the discussion that follows, they apply to a high degree in police and corrections officer subculture, especially when we consider maximum security institutions. They would also apply, to a lesser degree, in medium and minimum security facilities.

What did the individuals I interviewed have to say in general about subculture and subcultural constraints?

The officer subculture was defined in various ways. One respondent, for example, talked about the subculture as an unwritten but organized set of values that are communicated among the staff. Others talked about the code as an informal one, but one which, nonetheless, is known by all officers. Despite its informality, and despite the fact that it is unwritten, the code is often thought of as having a strong influence on the behavior of officers. One experienced administrator, who had himself come up through the ranks, described the subculture as a force that has the potential to make a "major impact on the judgment of officers."

The Importance of Remaining Solid

The loyalty or solidarity issue seemed to loom large in many of the responses. One justice educator, in describing subculture in the corrections context, said, "It seems to me that what you get is that same 'protect your own self and your partners' mentality that I see in other areas of law enforcement . . . where regardless of how serious the violation of either a code of ethics or a violation of policy, people have convinced themselves it is wrong to 'rat.' " Another educator talked in terms of there existing a very strong expectation to remain "solid." Some people talked about the loyalty or solidarity requirement as though it were part of the unwritten code. "I believe," said one person, "there's a code among officers—you don't put it on a fellow officer. If you go outside the norm and inform, I think you'll have a great deal of difficulty." Another said, "The downside is that there is a tremendous amount of pressure to conform with one's peers in ways that could put one at great, I'd say ethical, and sometimes even in legal jeopardy."

What is this pressure, and what might be the consequences of resisting it?

One educator, a former officer, said that "if you choose as an officer not to conform, you would be brought under a significant amount of social control within a very, very brief period of time. I have seen situations where people have had to deal with crisis situations on their own because they are now perceived as a loner, and as somebody the rest of the officers don't want to have anything to do with, because they are not conforming."

The officers invariably confirmed this idea that there would be considerable risk of social isolation for a colleague labeled as a "rat." "If you rat on somebody, you tend to lose a lot of friends, and, basically, you are left with no one to turn to for anything," said one. Another officer said, "You wouldn't want to 'rat out' on a fellow officer. It is something you just wouldn't do. You would be labeled a 'rat.' " I asked another person what the consequences might be, if any, for any officer reporting a serious infraction of a colleague. "I believe I would be shunned. I would be given a hard time. I would be an outcast. I would be labeled a rat." This shunning, as this officer put it, was often referred to in the interviews as "being put on the grease." For example, one officer said, "You put them on the grease. You don't talk to them, you don't sit with them. You treat them like outcasts."

The rule of thumb—the basic expectation—said another officer, was that "you keep your mouth shut and do not rat other people out." He indicated that the rule was that "if you saw a colleague doing something wrong, that was his/her business." As a brief aside here, one of the more disturbing aspects of working with justice students prior to, and after completion of, their field practicum is that one sees a value shift in individual students from an "I don't think I could let that go" (about a serious officer indiscretion) attitude prior to placement to a "so long as I don't do it—what she does is her business" attitude afterward.

A trainee police officer recounted to me some advice given to him during his first few days at a police unit. "The first day I was there, you know, after having met me and that type of thing, they [the officers] took me in their office and basically had a chat with me. They said whatever you see here, you keep your mouth shut you don't say nothing [sic]. I kind of looked at that two ways. You know, you see a lot of stuff the general public shouldn't be told about, but also maybe there was a kind of double meaning there. Maybe they were also telling me, whatever you and the other cops are up to out there, don't be going and squealing to his boss or something."

One corrections officer in a maximum security institution went so far as to say that it would be virtually impossible for someone declared to be a rat to survive in that facility. He also indicated that it would be equally unlikely that such an officer would survive in another institution because the officer grapevine, much like the inmate grapevine, would send word to that institution that this staff member was not to be trusted. Of an officer declared to be a rat, he said: "There is no way you can last more than a month or two months with the type of abuse you would take in this institution." He said the abuse would probably take the form of the "rat" being treated with "dead silence," having letters mailed to him with "rat" on the address, having her tires slashed, or "whatever." When asked to elaborate on what "whatever" may have signified, the officer declined to do so.

Report Writing

To draw this section of the chapter to a close, another officer was direct about what expectations the officers have of each other. "Say a situation happened and something didn't go down the way it was supposed to, and an officer is in the wrong. You don't bring this to anyone's attention, you stick together as a unit and you write reports as a unit." An administrator said something similar: "It is not unusual to have people sit down collectively and write reports. The individual who deviates away from his peers runs the risk of ostracization. They run the risk of outright intimidation from time to time."

What can also happen—and this is a fairly common theme in the interviews—is that if you see a wrongdoing by a colleague and you are questioned about it, you will simply respond that you did not see anything. What is implied here is that there may be times when individual officers are required to write reports or respond in a way they know to be patently false. They know that what they are writing is essentially a lie and they may feel bad about it, but the expectation is for them to become part of an event that is unethical and/or illegal—or face the consequences.

If you think about it, these are difficult circumstances under which to make personal, individual moral decisions.

I don't say this by way of judging you or your colleagues. I feel for you as you wrestle with the fine tension between the basic human (and professional) need you have to be accepted by your colleague group, and the need for you to be able to do what you consider to be the right thing in a given situation. I know your struggle. I have been there. I don't set myself up here as a paragon of virtue, or as someone having a monopoly on wisdom. I raise the issue of subcultural constraints and individual moral decision making because I think it should be raised, as difficult as it is to do so. But I raise the issue in a spirit of compassion, not of judgment.

It is important at this juncture that we add some riders. It would be wrong for anyone to suggest that there are no individual officers who have managed to achieve a status with their colleagues such that they can take their own moral stands and survive in their groups. Generally, these are officers who are experienced and respected as professionals; they are people who have proven themselves in dangerous situations and are seen as colleagues who can be counted on in a crisis. On occasion these individuals can, and do, take risks with their group that younger officers dare not take.

It would also be wrong not to add that there are sometimes significant individual differences even among the younger officers. Some are stronger than others, and are able to take more courageous stands, despite the subcultural pressures against doing so.

A Possible Gender Difference

Another interesting idea that came out of my research—and this has been confirmed in several conversations since—is that there may be a gender difference with respect to the personal stands that officers are prepared to take. I hear that women are more likely to take stands. I also hear that this may be due to the fact that they are not subject to the potential for "parking lot justice" that males may sometimes expect. It would be quite inappropriate, of course, to place too much stock in this idea at this time. And we should beware of making gross generalizations based on some personal observations. The idea of gender difference and moral courage may, however, be an interesting topic for further exploration in your classes.

In the meantime, here's some food for thought on this issue. In his autobiography *Moab Is My Washpot,* Stephen Fry, the eccentric British comedian and author, refers to a critique of a book called *Abinger Harvest* by a critic called Desmond MacCarthy. In talking about E. M. Foster, the famous author, MacCarthy says:

> Now, the essentially masculine way of taking life is to handle it departmentally. A man says to himself: there is my home and private life of personal relations; there is my business, my work; there is my life as a citizen. In each department he has principles according to which situations can be handled as they arise. But in each department these are different. His art of life is to *disconnect;* it simplifies problems. . . . The feminine impulse, on the other hand, whether on account of women's education or her fundamental nature, is to see life more as a continuum.

Is this why male officers may often seem to check their street morality at the front door of the institution or police agency when they go to work? Many male officers have said to me that this is how they learn to cope with the stress of their work. They compartmentalize their morality and operate by a different set of rules while on the job. According to MacCarthy, men are acting true to their nature when they do this. Women, on the other hand, are less likely to operate in this way. Because their approach to life, including their morality, is viewed more as a continuum, perhaps we should not be surprised at the way in which they are more likely to set limits, because this behavior is simply an extension of who they are by nature. This means that the rules of decency and propriety they live with on the street are more likely to be part of their professional performance.

Some interesting work has been done on gender differences and ethics among police officers in Illinois (Statistical Analysis Center, Illinois Criminal Justice Information Authority, 1994). It would be inappropriate to rush to sweeping conclusions based on the research but it does seem to indicate that female officers carry to their work as officers a less compromised ethic. For example, in response to the question "Over the past twelve months have you personally observed a fellow officer do the following: Use excessive force?, Cover up?, Fail to report excessive force?," a larger percentage of female officers in each case reported in the affirmative. This was also the case with respect to another question asked: "Over the past twelve months have you personally observed a fellow officer do the following: Stop/frisk to harass?, Give false testimony?, Write a false arrest report?" A greater percentage of the female officers reported that they had witnessed these wrongdoings. In other words, it would seem the female officers were more willing to be honest about their experience. They tend to identify problems more honestly than their male colleagues.

Dr. Gary Sykes, director of the Center for Law Enforcement Ethics, surmises that the reason for this higher level of sensitivity to issues of right and wrong might be what he calls "feminine morality," that is, that women in all societies seem to have a greater sensitivity to issues of care. As a result, they appear to commit more to a "rehabilitative ethic" than a "just desserts ethic," to what he calls a "peacemaking justice." If this is indeed the case we can see how women's tolerance for wrongdoing may be lower than that of males.

Of course, a word of caution is important here. We must be careful not to make general assumptions nor jump to conclusions. Most male officers operate in an exemplary way at work, and some women do not. This discussion is offered to get your intellectual juices going. The concept that men are more able and willing to compartmentalize their lives and their morality is, I think, an intriguing one and it may have a ring of truth to it. However, and I speak as a man here, it is difficult for us males to face issues like this one dispassionately; there may be a real temptation to react emotionally.

As a matter of fact, that topic is being explored in Mexico City as this chapter is being written. In a city noted for traffic, pollution, and police corruption, the chief of police has taken the highly controversial step of stripping his male police officers of the power to write parking and traffic tickets, giving that duty, instead, to teams of female officers. In his view, the only way to break away from the historical and deeply entrenched system that permitted male officers to augment their monthly salaries by accepting bribes is to replace them with women. It will be interesting to see how this experiment plays out, for even some womens' rights advocates in Mexico have pointed out that the only difference between male and female officers is that, traditionally, females have not had ready access to power like their male counterparts.

Backup

One final note. In some of the conversations I have had with officers and administrators, reference has been made to the idea that an ostracized officer could not be sure of backup in an emergency. It needs to be pointed out—and this, too, has been confirmed for me in less formal conversations with various individuals—that even someone who has been declared a "rat" will almost invariably be afforded colleague support in a crisis situation.

Reflections

What do you think of the comments of the police and corrections officers included in this section of the chapter? If you agree with all or some of them, would you be prepared to express them to

a. a researcher you did not know?
b. your colleagues?
c. your manager?
d. your teacher?

A number of key terms relating to officer subculture are listed in the preceding section. Which ones, if any, would you agree are significant in developing a better understanding of the norms governing your colleague group?

Why is it, do you think, that an unwritten, informal code of conduct can potentially have such a powerful influence over the decision making and behavior of individual officers?

One officer said: "There is a tremendous amount of pressure to conform with one's peers in ways that could put one at great ethical, and sometimes even legal jeopardy." How would you respond to this statement?

Should there be a place for discussion of officer subculture in your training or education? If so, why? If not, why not?

In your view or experience, are female officers more or less likely than males to take a stand on moral issues that relate to corrections or policing?

What is your view of the idea that men are more likely to compartmentalize their morality than women?

Let us now look at what some of your colleagues and managers have to say about the potential links between the subculture of officers and that of their "clients."

Links Between Officer Subculture and Criminal Subculture

This section applies principally to corrections officers, but there are pieces that may apply to police officers. In the case of corrections officers, of course, we are talking about imprisoned offenders, and in the case of police officers we would be thinking more of the "street" criminal subculture.

Shared Values

It's a troubling idea, I suppose, that there may be some similarities between the subcultural norms that govern the conduct of those in the criminal world and those that can sometimes govern the conduct of officers. But the links appear to be there, nonetheless. This was a common theme in the interviews I conducted. Consider how Vernon Fox (1983) in *Correctional Institutions* describes prisoner subculture, and think about whether there may be some parallels with the way police and corrections officers govern each other's conduct at work: "The inmate code in a maximum security prison is made up of the customs and folkways by which the residents protect themselves from the repressive measures of the administration; the inmates have nothing to do with the administration or correctional officers, and they do not 'squeal' or 'snitch' on fellow inmates. This inmate code, which is fundamental to prison organization, manifests itself in a type of informal control . . ." (p. 101).

Perhaps it is not surprising, then, that one corrections officer I interviewed said that in his view, the officer subculture is defined by the subculture of what he called the "clients we serve." The officer said: "So a lot of the subculture that you would see among corrections officers is very defined by the inmate subculture." He added: "One doesn't have to work long in institutions to find that people take on the manners of the inmates and their culture."

You will know, whether you are a police officer or a corrections officer, or a student or trainee preparing for either one of these professions, that there is a strong code of conduct that governs the behavior of the criminal element on the street, as it does the conduct of inmates in institutions. As we noted earlier, any criminal person not considered "solid"—where "solid" means, in essence, fully committed to the group and its code of conduct, including a code of silence—lives dangerously. In an institution, an individual identified as not "solid" often needs to be separated for his own protection. On the street, police informants live equally dangerously. I think it is clear that this value of absolute loyalty to the group can and often does govern officer conduct.

Which Came First: The Chicken or the Egg?

Where this de-individuating value has its roots is a moot point. I call it a de-individuating value because it requires the individual to sacrifice individuality in the interests of the collective. Did this value start with the officers, such that the criminal element adopted the value for itself for its own protection, or was it the other way around? One person I interviewed saw it this way: "I think there is a corrections officer subculture that originates or takes its origins from the inmate subculture. I think it is just as devious and I think it has as much of a dark side as the inmate subculture." Another person said: "So I think it is a learned form of behavior [the requirement not to 'rat'] that isn't attributed in the system to the inmates, but in my estimation that is where the staff pick it up. They start to play a similar role in how they interact with each other to the way the inmates do."

Solidarity

Wherever the answer concerning the origin of the loyalty norm, one thing seems clear. If you have that sense of solidarity on one side of the fence, you are probably going to have it on the other side. It is, after all, a very human way to react.

It probably has something to do with presenting a united, mutually protecting front against a common "enemy." One corrections officer expressed this idea well when he said: "I think the inmate code influences us in the fact that you have two distinct groups. You have got the inmates and they will do whatever they can to get by you, and they will stick together no matter what. And we have the same idea; if we are going to have any luck in controlling this place we are going to have to stick together too."

If what we have discussed so far is at least part of the dynamic constituting criminal and officer subculture, then we can see how the breaking of ranks by a member of either the officer or offender group is not going to elicit a positive response from that member's group.

So what did the practitioners who were interviewed have to say about all this? Not surprisingly, they offered some interesting and, I would have to say, sobering ideas. We don't like to think of ourselves as sharing *any* values with offenders, do we? But in fact we probably do.

By way of introducing their ideas, Mark Baker (1985), in his book *Cops: Their Lives in Their Own Words,* quotes a police officer as saying: "The mentality [of police officers] is an organized crime mentality. It is to be respected in some ways, but it has to be respected from the perspective of criminality, not from legitimate law enforcement. The culture that we come from and the kind of country that we live in, we do respect that kind of camaraderie. But it's magnified in police work. It can go too far" (p. 175).

May I suggest that Baker, even though he does not elaborate on the point, is suggesting by the phrase "It can go too far" that carrying the norms of loyalty and solidarity in police work too far can and sometimes does result in unethical conduct on the part of some individual officers. This sometimes blind allegiance to the loyalty norm can create fertile ground for officer behavior ranging from poor judgment at one end of the spectrum to corruption at the other end.

For anyone involved in the "teaching" of ethics, the issue of accountability is an important one. In the process of changing the culture of a police organization for the better, a number of different avenues are available for exploration. The chief or sheriff may choose to stand up in front of her organization and talk about the values of the agency; a consultant may be brought in to conduct a survey and provide training; an organizational newsletter highlighting and discussing ethics issues could be created and distributed across the department. The single most effective demonstration of ethical soundness in any organization, though, will be a ready willingness by all employees to hold *each other* accountable for their actions.

In February 1999, four members of the New York City Police Street Crime Unit fired 41 times at an unarmed civilian (striking him 19 times) in the vestibule of his own home. Several factors combined to make this a particularly inflammatory event. First, the innocent citizen was black and all the officers were white. Second, this particular unit of the NYPD had a reputation for aggressive and confrontational tactics. Third, this event came on the heels of several other highly publicized and widely criticized interactions between the police and the minority community in New York. As a result, a number of private citizens, celebrities, retired police officers, and elected officials engaged in protest demonstrations and acts of civil disobedience in front of the NYPD headquarters. One who chose to submit to arrest was New York Congressman Charles Rangel. As he was being taken into custody, he said he would like to hear "the decent men and women of the Police Department, who are in the vast majority, shatter the blue wall of silence" and speak out against misconduct. "I want to make a special appeal," said Mr. Rangel, "to the tens of thousands of courageous police officers who work hard on our behalf every day not to allow this contamination of their department by the few. When these good officers see wrongdoing, we need to hear them say that they won't tolerate it" (*The New York Times,* March 16, 1999).

Lynne Samuels, a New York City broadcaster, was interviewed by Michael Enright on *This Morning,* a program aired in Canada by the Canadian Broadcasting Company. The interview took place during the week the grand jury was hearing evidence in this case. Ms. Samuels, a white New Yorker, made the point that the current, "Quality of Life" campaign was being conducted at a significant cost—a cost paid mainly by New York City's black population who were systematically being subjected to having their rights and freedoms infringed on by some police.

Then Ms. Samuels made an interesting point. She said that the injustices committed by the police were not being exposed as they should because of the way police officers were being governed by the requirement for them to remain solid. She said: "Let's say that only about five percent of the forty thousand New York City police officers are bad, that's two thousand of them running around our city every day doing their thing. The problem is that none of the thirty-eight thousand other officers will say anything." Michael Enright added: "They probably remember Frank Serpico from the seventies and what happened to him when he broke ranks." Serpico, you may remember, was a New York City police officer who attempted to expose corruption on the force. He paid a terrible price for his attempt. He ultimately got shot in the face and is convinced that he was set up by his colleagues. The solidarity norm would appear to be a very compelling one indeed, at least on the New York City police force if not elsewhere.

Another graphic example of this blind allegiance took place in Toronto. Two Metropolitan Toronto police officers were charged with planting drugs on a suspect whom they subsequently charged with possession. In their trial, evidence was given by a young, 26-year-old police officer, who had been convicted (and dismissed from the force) earlier for attempting to obstruct justice in the case by falsifying his notebook entries. In a newspaper account of the court proceedings, the young officer offered the following in his defence: "I went along with the flow. I went ahead with it. . . . I didn't want to rock the boat." He said he was very uncomfortable with his actions but continued in this way to justify them: "I wouldn't have much of a career left as far as people working with me. . . . You're supposed to be a team player" (*The Globe and Mail,* October 23, 1996, p. A9).

Maureen Prinsloo, chair of the Metropolitan Toronto Police Board, made a decision to tour some police stations after the two accused officers in the above case were found not guilty of planting drugs on the suspect. In a newspaper account of her visits, Prinsloo is quoted as saying "the vast majority of officers simply did their jobs." The article continues: "However, she said, when she asked 'Why do you turn a blind eye to this small group who will pull your reputation down?' I got no answer." "To Serve and Protect (the force's motto)," she said, "that is to serve and protect the public. It does not mean protecting the bad ones in your midst" (*The Toronto Star,* November 21, 1996, p. A6).

In 1999, a 19-year veteran former police officer on the Metropolitan Toronto Police Force was charged with a series of drug trafficking offenses involving high school children. This is an officer who, it turns out, had been in trouble with the police brass on many occasions before. His penchant was to use unorthodox methods to "get his man," many of them illegal. He was seen by his colleagues to be slippery and untrustworthy but it wasn't until he got involved in the drug dealing that his former colleagues were prepared to say anything about their former colleague.

Christie Blatchford who often writes on justice matters for *The National Post,* a Canadian national newspaper, said this about the situation: ". . . it was only when the drug trafficking allegations came to light did many of his former colleagues consider that he'd crossed the line, which provides an interesting glimpse into the subculture of even one of Canada's most progressive police forces at the dawn of a new century: Roughing up prisoners may be okay; framing suspects is arguably tolerable if they're believed to be guilty anyway, but dealing heroin to young people? Well, that's beyond the pale" (*The National Post,* February 9, 1999, p. A1).

Here's the case of a police officer who had survived for many years even though many of his colleagues did not trust him; they knew full well that he was given to, let's say, "unorthodox" (read illegal) policing methods. But it wasn't until he had crossed way over the line, and, indeed, was no longer one of them, that they were prepared to break ranks, breach their code of silence, and overcome their commitment to remaining solid. Indeed, you may well be correct if you are thinking that this does not amount to breaking of rank because the officer was no longer one of them—he had gone "bad"; he was a rogue officer.

(As an aside here, one of my research findings points to the idea that individual officers have differing "lines of tolerance" for colleague wrongdoing. However, one commonly held idea seems to be that officers fraternizing with the criminal element or willfully engaging in lawlessness generally will be held accountable.)

One haunting thought for me in this case is this: What if this officer had been held accountable by his colleagues earlier in his career? Might he have avoided hitting the painful jackpot he subsequently hit? I remember a few years ago reading a letter from the wife of a police officer in an Ann Landers column (no smart remarks, please!!!). The police officer had just been fired. He had done something terrible as a direct result of his alcoholism and his career came to a crashing halt. His wife wrote in her letter something to this effect: "My husband's worst enemies were all those colleagues who covered for him. Why didn't any of them hold him accountable?"

It's a sobering thought, isn't it?

One justice educator, an experienced corrections practitioner, indicated what he considers to be the cardinal norm in both officer and criminal subcultures. He said, "An inmate doesn't inform on an inmate, and an officer doesn't inform on an officer." And another officer said: "Just like the inmates have an unwritten rule that they don't rat out on each other, so it's an unwritten rule that officers don't rat out on each other. . . . Our code and theirs is basically the same code."

Another educator, a former officer and administrator, took this theme further. "But I also think of the unethical practices with respect to the subculture. I think the prisoner subculture has rules like 'you don't rat.'. . . Well, a similar process happens for correctional officers where they will be under pressure not to say anything if one of their coworkers does something that is considered to be against the rules. For example, hits an inmate, beats an inmate up, steals from an inmate, misrepresents the facts in a report. . . . New staff may be pressured to learn very quickly an opposite way of doing things from the way in which they have been taught. You join the system. But you get that way, you pick it up—in the same way police officers do."

As public fury swept New York City with the news that a Haitian immigrant had been tortured at the hands of the police, both Mayor Rudolph Giuliani and Police Commissioner Howard Safir publicly expressed their outrage. In his remarks, Giuliani spoke about his intention to "break down the blue wall of silence once and for all." He then went on to express his utter amazement that "officers who, themselves, would never consider doing such a despicable thing, will not step forward to inform on one of their fellow officers who has."

It's pertinent that a connection is being made between the subcultural pressures one can feel as an officer and what this person calls "unethical practices." It is critically important that we refrain from making any overgeneralized or *carte blanche* statements about the possibility of this connection, but it is a question that deserves serious and open-minded attention.

More Shared Values

Some of the officers interviewed talked about other shared values between the officer and inmate subcultures. One institution administrator, for example, talked about how both the inmates and the officers see the sex offender as on the lowest rung of the ladder in the inmate hierarchy.

Others talked about the way language is shared by the two subcultures. One officer said of his colleagues: "I think they pick it [the language] up from the inmate subculture. It is a strange thing but they pick up the inmate jargon. They talk about 'don't crack to me,' an inmate expression which means don't talk to me. Both groups use expressions like 'put on the dummy,' 'cheese eaters,' 'laying track,' that kind of thing." Those who have worked in the justice system will also be aware of expressions like "bit" (sentence), "diddlers" (child molesters), " 'deuce less" (two years less a day), "six" (officer coming), "the hole" (segregation), or "queen" (effeminate male inmate who may or may not be in the process of a sex change), to name just a few terms often used interchangeably by prisoners and officers. Many of these terms, of course, apply principally in the context of correctional institutions.

Police officers also share a common language with the criminal subculture. They, like corrections officers, would use terms like "rat," "fink" or "narc," and "diddlers." In the United States, police officers talk about "stand-up guys" (officers who will protect a colleague who has done something wrong), and "hand-up guys" (officers who will do the opposite). The terms refer, essentially, to officers who are considered solid or not solid. Police officers may also use terms like "screw" (correctional officer), "joint" (jail), "hopped up" (under the influence of drugs), "roach" (the butt of a marijuana cigarette), and "score" (to purchase drugs). Mark Baker (1985) in *Cops* quotes one police officer as telling him: "So many police terms are picked up from the street world and become incorporated in police mentality [and language]" (p. 199).

Who influences whom in the use of this common argot? I put this question to an institution administrator who stated very clearly his view that it was the inmate language that influenced the language of the officers. He said: "You don't have inmates going around talking like corrections officers or acting like them in any way, shape, or form."

Reflections

Would you agree there may be some links between officer subculture and criminal subculture?

If you agree, what would some of these links be? Can you think of some of your own?

Do you agree that officers and the criminal group can come to share a jargon?

Subculture and "Enemies"

We have suggested that subcultures are stronger and may require for their existence an identifiable common enemy. In the case of police and corrections officer subculture we have suggested that the "enemy" may take three forms. First, the criminal element; second, the administration; and, third, the public.

It should be pointed out that the primary target for officers' antipathy is the offender group, followed in order by the administration, and the public. This is summed up nicely by Mark Baker (1985), also in *Cops,* when he talks about police officers' attitudes toward these groups. "Many cops begin to see themselves as under siege from all sides. They develop a classic 'Us versus Them' mentality. . . . The real reason most police officers socialize exclusively with other police officers is that they just don't trust the people they police—which is everybody who is not a cop. . . . Older officers teach younger ones that it is best to avoid civilians. . . . The 'Us against Them' attitude also applies within the ranks of the police themselves. Superior officers from sergeants on up are regarded by patrol officers as 'Them' " (pp. 175–176).

The depth, breadth and rigidity of the police subculture often surprises those outside of it. It affects the manner in which law enforcement families interact, and sometimes extends to the very formation and geographical development of communities. In *My Father's Gun* (Dutton, 1999), Brian McDonald wrote about growing up as the son, grandson, and brother of police officers, and described how even on vacation, his family would be "surrounded by cop families." In the 1960s, the New York Police Camp in the Catskill Mountains could accommodate 100 cop families, and there would always be a waiting list for space in one of the bungalows or the main hotel. Even the waiters at the camp were sons of police officers (many going on to join the force themselves), including one who even became the police commissioner—Robert J. McGuire—who had worked at the camp in the 1950s.

According to McDonald, the New York City Police Department was, in the late 1950s, the "most insular of societies." As a largely homogeneous (Irish Catholic) paramilitary organization, the notion of "brotherhood" was enhanced by the danger of the job and a common enemy (criminals). When the emigration to suburban areas like Rockland County, New York, began, it was only natural that police families would cluster together. McDonald described things this way:

> They car-pooled and socialized together. They joined fraternal organizations like the Knights of Columbus. Their families—our families—went on vacations together to the Police Camp in the Catskills. For city cops, there was no good reason to venture outside their circle. In their minds, the outside world was a place they had little in common with, a place that did not operate under the same rules and codes.

Anti-Criminal

It should be acknowledged that there are two scholars who, in the context of corrections, argue that corrections officers are not as anti-inmate in their disposition as they themselves, and others, sometimes think they are. John Klofas and Hans Toch (1982) developed the concept of *pluralistic ignorance* among corrections officers. Based on a survey of corrections officers, they are of the opinion that there are what they call "progressive" officers (officers who are not anti-inmate) who, thinking they are in the minority, are, in fact, in the majority. Likewise, officers with nonprogressive (anti-inmate) views, feeling themselves to be in the majority, are, in fact, in the minority.

The results and conclusions of their survey are curious. My research and experience present a different picture—that officers in the main are, or become, anti-inmate in disposition. One or two experiences of being "conned" or manipulated by an inmate can quickly change the positive officer into a suspicious one who sees the inmate group as the enemy. And generally speaking, any officer who continues to be essentially pro-inmate in the face of such experiences is often viewed with great suspicion by his officer colleagues.

One administrator I interviewed said: "I know what the code is and it is very strong. A person [new officer] can come into this work with all good intentions, but a lot of people just give up and fall into the anti-inmate routine because it is easier—make your money, go home, and forget about it." And another officer talked about how an officer's attitude toward inmates can change: "It's a funny thing.

When you first start here, you tend to sympathize and empathize with the inmates, but after dealing with them on a daily basis, when you are treated badly, in a way you are almost happy if something does happen to them."

One officer made a stronger statement when he talked about officers having "an inherent dislike for inmates, not a feeling of revulsion necessarily, but certainly a dislike for most of the inmate population just simply because they are inmates."

In the case of police officers, the case is an even more clear-cut one. There is a strong antipathy toward those who break the law, and who represent, potentially, personal danger for the officer. In a chapter of *Cops* entitled "Blood Brothers," Baker (1985) talks about how this antipathy toward lawbreakers and the danger they represent can galvanize police officers into a solid group. "Nobody understands but another cop," one officer is quoted as saying.

Let's take a brief look at what people in my interviews had to say about these antipathies.

One clinical psychologist who works in the justice field talked about how the relationship between officer and offender is characterized by an "us" against "them" attitude, and he attributed the binding together of both officer and inmate groups to this way of thinking and feeling. An experienced administrator and former officer had this to say about institution subculture: "It is 'them' against 'us,' 'them' being the inmates or the administration, depending on the issue." An officer expressed some strong anti-inmate sentiments: "You are dealing with people who go against every principle you have in your life. You are on the other side. . . . You have got them treating you like you are a lowlife." Another officer was even more strident in his account: "The minute you walk through the door, you are in a whole different environment. . . . If the inmates can get something off you they will lie through their teeth to get it. . . . You get burned a couple of times by inmates and, yeah, you are going to come out with a very cynical attitude—and that is the attitude you have in this place." One female officer, equally strident, said: "The general public has no idea what kind of animals society has locked up, and you are dealing with these people on a daily basis. . . . You are no longer part of a polite society now you are in amongst animals. . . ."

Police officers express similar sentiments about their professional lives and the kind of people they have to encounter during the course of their duties.

Anti-Administration

The anti-administration feelings police and corrections officers can develop is illustrated by an event that took place in the Metropolitan Toronto Police Force in

January 1995. A one-day illegal strike of police officers took place in 51 Division. The deputy police chief had decided to order a public inquiry into the conduct of two officers from that division, and this decision precipitated the strike. One of the columnists in *The Toronto Star* (January 30, 1995, p. A6) wrote this about the incident: "When you try to penetrate the miasma of cop culture, this is what happens. When you cross the line of blind allegiance to cop fraternity, this is what occurs. Politicians are given an ultimatum. The public is threatened with the withdrawal of vital services. And the full ferocity of cop wrath descends on the head of the deputy chief who was already seen—because of his commitment to accountability and procedural reform—as the enemy."

A corrections officer summed up this anti-administration idea succinctly when she said: "Because the management is after us, you cannot have everybody running around totally against each other." And another officer resorted to the use of inmate jargon to explain where he stood on this issue: "You don't talk to them [the management], you don't 'crack' to them, only in the line of business. You just say the minimal amount that you have to say to them to get the job done."

In the United States, police employee groups have begun to use the "no confidence" vote as one means of publicly expressing dissatisfaction with chiefs and other top managers. Typically the leadership of a union or association will poll the membership on the competence (some would say popularity) of the chief, after which the results are released in a press conference. While having no formal power, there is a very real political element; no mayor or city manager, after all, can avoid responding to such a contentious public statement. Anthony Bouza is one police chief who led several different agencies in his career and was not without his critics as he did so. Addressing a leadership forum several years ago, Bouza spoke about the difficulty he faced in trying to bring about change in the management of police organizations and the resistance he often encountered from various employee groups in the process. In discussing his experiences in the Minneapolis, Minnesota, Police Department, he described the results this way: "When I arrived, the agency was badly divided within itself. By the time I left, I had brought them all together. They all hated me."

Anti-Public

Finally, here are two sentiments, one expressed by a corrections educator, the other by a police educator who had worked as a police officer for many years. The former talked about what he called the "hunker down" mentality of officers. To explain what he meant by hunker down, he talked about "inmate against guard, guard against inmate, guard against management, guard against the collective community." He added: "Where they, correctional officers and police officers, feel they are misunderstood and that their work is too difficult, and too dangerous, and 'because no-one loves them', what they do is band together and they hide themselves behind the culture of not talking about it, because most of us [the public] don't understand."

The police educator talked in similar terms about police officer subculture. He talked about what he calls the "musk ox syndrome": "You know about musk ox. They are large northern animals in the buffalo family. It's interesting, when they perceive themselves to be threatened they form a circle with the cows and the calves in the center of the ring and all the bulls stand shoulder to shoulder with their massive heads out. And, oftentimes what happens is that the subculture within the police is the tendency to pull into a ring similar to that."

The former police officer continued to offer an explanation for this musk ox syndrome. The attitude of police officers, he said, is best characterized in this way: "The only people we can really trust, the only people we are really comfortable with is ourselves—other members of the subculture. So therefore I should pull into the subculture and don't blame me if I don't talk to you." In expanding further on police attitudes toward the public, he added: "We don't want you to like us, we don't want you to know, or we don't want to know that you like us because it doesn't allow us to pull back into the circle so readily, so the more I can push you further away, the more comfortable I feel with being part of the subculture."

This anti-public sentiment is summed up by something a very experienced corrections officer recounted to me. "This job can really screw up your thinking about things and about people," he said. "When you are first in the job and you see someone drive into your driveway at home you say or think to yourself, 'Who's that in my driveway?' After a few years you see someone drive into the driveway and you say, 'Who the hell is coming into my driveway?' After several years in the job you see a person coming up your driveway and you say or think, 'Who's that scumbag coming up my driveway.' " Those officers whose view of the general public deteriorates in this way don't like what they see themselves becoming, but seem to be powerless to stop themselves from developing this jaundiced view of folk other than their colleagues.

This kind of thinking is confirmed by researcher into police behavior Jerome Skolnick (1975). He suggests that the authority police have over the public, and the danger they feel themselves to be in, causes the police to isolate themselves from the community and to look to one another for support (p. 44).

Finally, Vernon Fox (1983) in his book *Correctional Institutions* offers a similar perspective with respect to the corrections context: "The staff tend to develop loyalty to 'prison people' to insulate them from intrusion or criticism from the outside community. . ." (p. 162).

Summary

This chapter continued to explore the discussion on officer subculture that we started in Chapter 5, by incorporating and considering the perspectives of correctional and police practitioners.

Their perspectives confirm the idea of officer subculture as one characterized by an unwritten, informal code of conduct influencing, if not governing, the conduct of officers at work. Chief among the norms of the code, as is the case with the criminal code, is the requirement to remain loyal to one's colleagues. The practitioners all talked about the probable and unpleasant consequences of disloyalty.

Links between officer and criminal subculture were confirmed. In addition to the loyalty norm, both groups tend to see the sex offender on the lowest rung of the offender ladder. A shared argot also appears to exist.

The practitioners confirmed that the officer subculture requires identified "enemies" for its continuing existence. These enemies were identified as the criminal element, the "bosses," and the public.

Reflections

Would you agree that members of a subculture are likely to bind together more closely if they have identified "enemies"?

What are your views on the discussion in this chapter with respect to officer subculture being anti-inmate, anti-administration and anti-public?

Is it important that officers hold each other accountable for their behavior?

What techniques can you think of that may help you keep a sense of perspective about members of the general public, that is, that you don't begin to see virtually everyone other than your colleagues in negative terms.

Tough Decisions

The way I see it, the bottom line for making decisions when there is peer pressure, is to ensure that you are happy with the decision for yourself, not for others.

—Journal entry, justice student in ethics course

When making an ethical choice, one must consider the truth versus loyalty, the individual versus the community, short term versus long term, and justice versus mercy. The bottom line though is that you have to make your own decision because you have to live with it.

—Journal entry, criminal justice student in ethics course

Once we get into the field we will depend on the money, that is the nature of our society, and we will justify almost anything to maintain that paycheck. How will you explain to your wife and kids (not to mention the bank) that you quit your job because of an ethical or moral decision?

—Journal entry, criminal justice student in ethics course

To have the bravery to challenge one's convictions is necessary for the evolution of a person's psyche, soul, and personality. But the process is painful and we all tend to shy away from it.

—Journal entry, criminal justice student in ethics course

Chapter Objectives

After reading this chapter you should be able to

- *Explain how subcultural pressures can sometimes affect the choices officers make.*
- *Explain William Tafoya's concept of the* ***vortex.***
- *Explain how the roots of officer wrongdoing may reside in allegiance to the solidarity norm.*
- *Identify two possible ways in which an officer's moral choices can adversely affect his sense of personal and professional well-being.*

The Perils of Unthinking Loyalty

Permit me to stick my neck out a country mile (as if I haven't stuck my neck out already!). I suggest to you that the solidarity norm of police and corrections officer subculture constitutes fertile ground for officer wrongdoing. I am *not* suggesting that because police and corrections officer subculture is characterized, among other norms, by the solidarity norm, all officers are therefore wrongdoers. As we have suggested in earlier chapters, individual officers have found ways to maintain a sense of personal autonomy in spite of the subcultural norms. There is little doubt that most officers are honest and professional in the way they carry out their responsibilities.

What I am suggesting, however, is that it is not unusual, on those rare occasions when an individual officer or group of officers is adjudged to have acted unprofessionally, to discover that the solidarity constraint may have played some part in the initial wrongdoing, and almost certainly in any subsequent wrongdoing relating to the original incident.

Let me give you some police and corrections examples. We will look first at the policing context. Consider this example:

When he received the radio transmission about an auto accident along a rural highway, the young officer responded immediately. Driving up to the scene several minutes later, he saw a heavily damaged vehicle lying on the pavement, and an obviously distraught female running toward him from her hiding place behind a highway guard rail. When she approached, she told the officer that another car had crossed the center median of highway striking her vehicle, and that when she got out of her heavily damaged car, the driver of the other automobile (who appeared to be intoxicated) had tried to run her down. To avoid him, she ran behind the guard rail, after which he struck that metal barrier before fleeing the scene. Fortunately the woman was not seriously injured, and she had the presence of mind to take note of the license plate of the vehicle. That, coupled with a substantial amount of debris left by the hit-and-run vehicle, heartened the one-year veteran officer, for it seemed the culprit in this accident might be quickly apprehended. Using the license plate information supplied by the woman, the officer called his dispatcher to learn the identity of the owner, and was shocked at the reply. The last name of the vehicle owner was extremely unusual because of its unique spelling, and it was identical to the last name of one of the police officers with whom he worked.

When this information was broadcast, other officers heard it as well, and a senior sergeant working on that shift responded to the scene. Taking stock of the evidence and information, the sergeant ordered the young officer to write the accident report without including information about the license plate or other evidence, and told the youngster that he (the sergeant) would "take care of everything." To his everlasting shame, the young officer went along with the sergeant's plan.

To make a very long story very short, the driver of the hit-and-run vehicle was the brother of the police officer with the unusual last name and he had, in fact, been drunk when he had the accident and attempted to run down the woman motorist. Both the young officer and senior sergeant were arrested, and charged with a variety of offenses including the felonies of perjury and tampering with evidence. Using his relative inexperience and lack of rank as a defense, the young officer was able to negotiate a sentence of probation instead of incarceration. The sergeant was not as fortunate, and was sentenced to a term in jail. Both of them, of course, lost their jobs.

It is perhaps difficult to stand in judgment over the young officer. What is a colleague supposed to do? To act ethically and professionally in this case might well have caused him subsequently to be labeled as disloyal, as a "rat," as someone not to be trusted by colleague officers. He found himself between the proverbial rock and hard place.

One young woman recounted to me a disturbing account of a young police officer's experience. That police officer was her brother. "My brother is a relatively new police officer and he too has already felt the pressure of the officer subculture," she wrote. "He always said that if ever he became a cop that he would remain his own person, regardless of the pressures of the force. But even he, who is one of the strongest human beings I know, has fallen down when it came to making a decision with an experienced officer.

They were out roaming the streets," she continued, "when they got a call on the radio about a robbery that was going down in the 'projects' just minutes away from where they were. The old-timer said to my brother, 'I'm not racing to that location—there's nobody there but blacks—let someone else pick up the call.' My brother, who is clear of all racism, was shocked. He could not believe the officer was going to let the robbery happen because of the neighborhood in which it was taking place. My brother had a choice to let his supervisor know what this officer did (or in reality did not do) or to stay quiet about the whole thing. My brother was one of those students who tried and tried and tried to get on a police force and after many attempts he finally made it. He knew that if he ratted out his colleague to a higher authority he would be putting his job on the line—the job he had worked so hard for." Then she finished, "As much as my brother hated himself for doing it, he decided to stay quiet to save his credibility."

What a tough spot for this woman's brother to find himself in! The event proved to be a severe test for his character, but because of his fears engendered by the subculture he felt powerless to do what he thought he should do.

Another example is one that took place on a large metropolitan police force in Canada. Let me give you a little background. In 1996, a young police officer was tragically shot and killed by a drug dealer while on duty. It was a particularly sad day for the police department, not only because of the tragic circumstances, but because the young constable's dad was a retired detective with the same force. The sadness experienced by all members of the force, as you can imagine, was profound.

A year or so after the burial of the young officer, two police officers from the same force were out on patrol in an unmarked car when they noticed one of the occupants of a slowly moving vehicle expose his buttocks to three young women in another car. The cars were traveling along an inner city freeway. The car was stopped and the occupant doing the "mooning" was charged with performing an indecent act. The man exposing his buttocks turned out to be the brother of the deceased officer. The accused appeared in court and was given a conditional discharge, which meant he would not have a criminal record. He was fined $100 and ordered to write a letter of apology to the three young women (*The Toronto Star,* June 23–24, 1999, pp. A10, A22).

To cut a long story short, since the two officers laid the charge, they claim to have been harassed and discriminated against. One senior officer, they claim, said to them: "He's a policeman's son, for crying-out loud," meaning, presumably, that they should have responded to the situation somewhat differently. They have since been taken off plainclothes work and put back in uniform, something they see as punishment for what they considered to be a case of their applying the law fairly and impartially. As a result they submitted a 40-page report alleging corruption in the force.

This seems to be a case where two police officers appear to have paid a price for daring to break their adherence to the police fraternity and its code. Clearly, many of their colleagues and several of their senior managers felt the right thing for them to do would have been to issue a warning only—presumably because the offender was the brother of a slain officer and the son of a retired and well-respected veteran of the force.

What do you think the two officers should have done? What would the three young women have felt had the offender not been charged? What do you think they would have felt about the court's disposition? Is this one of those situations where discretion should have been applied? It's the old rock and hard place situation again, isn't it?

Mark Baker (1985) in *Cops* recounts an interview with a police officer who talks about how he and five other officers were called out to a domestic situation. "One kid moved over to one of the cops and said something to him. The cop pulled his gun, shot the kid in the head and killed him. Outright. Now you've got five cops standing there and going 'What did you do that for?'. . . They all got together and they figured, 'We got to do something. We'll say the kid had a gun or a knife or. . . .' They come up with all these stories. . . . Nevertheless, these cops felt that kind of bonding relationship with him to protect him. So they all lied. They didn't have to say that kid had a gun. But they wanted to protect their brother, because he was one of them" (pp. 199–200).

Here is a case of police officers, holding positions of public trust, being prepared to lie for each other even when by doing so they knew that justice would not be well served.

Here is one more example. It is not a "heavy-duty" example like the one above but I include it here because it speaks to how trainee police officers can be introduced to the subcultural norms. A police science student, after attending a police ethics seminar I had conducted, asked if he could speak with me privately. He said he was troubled. He had just completed a three-week practicum and his field training officer (FTO) was a 20-year veteran of the police force with which the student was placed. On one shift, he and the FTO were told to go and ask some questions of a particular community resident. When they got to the house they discovered that no one was home. The house was situated in the country and around it were several outbuildings. The officer decided to walk around the property and look in the outbuildings even though he neither had a search warrant, nor could justify obtaining one. They looked in one of the buildings and saw a van that they discovered was unlocked. The senior officer opened the door to take a look at the serial number and found that a piece of paper had been placed over it. The van had obviously been stolen. The senior officer said to the student: "Did you see a piece of paper over that serial number?" The question clearly was asked in such a way that the student was supposed to reply something like "What piece of paper?" In essence the officer was trespassing and acting illegally, and now he was going to lay a charge of theft against the resident. But it was important to him that he could rely on the student. He said: "When this gets to court it is very important that you back me up and you say that the serial number was visible."

What if this trainee had said that he would have difficulty doing that? What if he had questioned the actions of the veteran officer? One doesn't have to be a rocket scientist, I think, to recognize that the trainee is in a difficult situation here. If he assures the FTO that he can be counted on, he has in essence taken the first step toward—if not down—a slippery slope. If he indicates that he cannot make that promise to back up the officer in court, then he puts his future in policing in jeopardy because he will be seen as someone who is not solid and who, therefore, cannot be trusted.

(As an aside, there is a significant message here for senior officers when they select training officers, and for field training officers once they have been selected. Rookie officers emulate the behaviors of senior officers; they pick up quickly both the good habits and the bad ones. FTOs have a tremendous responsibility to perform at their best when being shadowed by a new officer. In a similar vein, police managers have a great responsibility to select trainers wisely because of their potential for influencing their protégés.)

Now let me give you some examples from corrections. The following account was given to me in one of my research interviews. I include the account in its entirety because it delivers a poignant message.

> This inmate was a real bad actor on this particular living unit where I was working at the [identity withheld] Detention Center and he ended up there. I watched him, and many of the staff had talked to me about him. I was the supervisor on the floor. And many of the staff over a period of days had talked to me about him and let me know that this guy was a trouble-maker. They couldn't catch him, but they knew he was causing trouble and creating a lot of tension.

He was picking on some of the black inmates; he was a white inmate himself, he was a big heavy guy, he was very strong, he was very muscular, and it looked like he controlled the unit.

One day I was walking past the corridor and I saw him cleaning outside the cells but in the day area where inmates would normally come out and watch television, play cards, and listen to the radio. All the other inmates were in their cells, but this particular officer had this guy out sweeping and mopping. This guy went to the cell of one of the black inmates, the inmate who was in his cell was no problem to us at all, he was very cooperative, doing his time, awaiting his trial, but he was also very big, about the same size as the other character I'm talking about. The cleaner went up to this guy's cell, the inmate inside the cell came to the grill and the cleaner—I watched the whole thing myself—spat at the black inmate on the inside, but there was plexiglass in between so it trickled down the glass. I immediately called for backup.

I had already made the decision this was inappropriate behavior, there was no cause for this from my observation, and I was moving him to segregation. You understand that I had lots of information prior to this. The staff had serious concerns about him and the tension caused by him. It was one example for me that I could use to put him in segregation, to take him out of the unit and maybe solve some of the tension. So I called for backup and about seven staff came to the living unit. I told them to wait around the corner and I went with one officer who unlocked the doors for me and I called this guy out, and he said, "Where am I going?" and I said, "I want to talk to you out here." I eventually coaxed him out into the corridor and he said, "Where am I going?" and I said, "You are going to segregation because I just saw you spit at that guy in his cell." He threw these big arms up and said, "There's no fucking way, I'm not going anywhere," and I thought he was going to hit me because his arms had come up and then I lunged with my hands to grab his arms and lean against him on the glass. Just then all these seven other people came flying around the corner and the dance was on for them. What they saw, it appeared to me, was an opportunity to take out all their historical anger from the last two or three days, and they were kicking him, they were punching him, and they were pulling his hair.

It was, in my estimation, a very clear case of assault versus restraint. All they had to do was restrain this man but the vented anger unloaded and it was all I could do as a supervisor to stop it. I was yelling and screaming, I was grabbing legs and pushing people back, and it was a very uncomfortable situation for me. This guy was beaten up and I got him to segregation, left him in his handcuffs and leg-irons for a little while before I would go in and talk to him, because he was pretty angry. He didn't report it. I didn't report it as a physical assault.

This officer then talked about how he felt about the situation and the role he had played in it:

> Was it unethical? Absolutely. Did I do anything about it? Absolutely not. He was placed on a misconduct and in his adjudication he didn't say anything about it, I didn't say anything about it, so the facts of the intervention never surfaced.
>
> This is an example of somebody—me in this case—of somebody who bought into the officer subculture and didn't want to get seven staff dragged through a carpet of suspensions and investigations. Now had the inmate said that he wanted to report it [the abuse] I think I would be here telling you a story about someone who had lied, who heard the inmate's allegations but went and defended himself and seven other people. It [the subculture] is a very powerful pressure.

Let me give you another example in the corrections context, details of which were relayed to me by a criminal justice educator who had worked for many years in correctional institutions. He said: "There was this situation where this inmate came in, was being admitted to the institution and this inmate was being very abusive to the officer, was very antagonistic okay, and the officer lost control. The officer assaulted the inmate. The officer came to me and said what he had done. And I said to him this is what you are going to write up." I asked the educator if the reports they wrote were fabricated. "It was a lie," he replied. He then said he had punched the officer to make it look like he had been attacked by the inmate. In this way the cuts and bruises on the inmate could be explained: There had been an altercation. "When I took that officer into the elevator I said 'You know what, we are going to make it look like this inmate attacked you. Close your eyes and I am going to punch you.' And that is exactly what I did."

In a young-offender correctional center, one new officer recounted the following experience. "A colleague was telling another staff member and me that he had banged a resident's head off the wall because the resident was being physically aggressive toward him. He had banged his head off the wall while trying to put him into a restraint. He had done it on purpose. He had documented it as an accident. The other officer was laughing and I was totally shocked and had a stunned look on my face. I also couldn't believe that the other staff had laughed about it. I couldn't say anything except ask him, 'What did the resident do when you knocked his head off the wall?' And he replied, 'He just cried.' That was all I said to him and my conscience is eating me up. I don't trust him around the residents while in rough situations because I know he would hurt them out of anger and would not care." There was a moment of quiet and then she said, "I would rather not say anything because I don't want to cause any trouble because I just got hired there. I know this is sad but I am telling the truth."

This is a tough situation for a new officer to find herself in, but it is not an entirely uncharacteristic experience in corrections. The officer in question made a decision not to respond to the event she had just experienced, although she made it clear to me she did not feel good about her decision. The constraint to fit in and be accepted was simply too strong for her at that point in her career.

While I was preparing the second edition of this book in late 1999, a college ethics teacher sent me an e-mail and the subject of it was listed as "Why we do this job. . . ." The teacher had been using the first edition of *Reputable Conduct* and had been discussing the book with a class of justice students. He had received an e-mail message from a member of the class and had forwarded it on to me. I sought and received permission from the student to include the message in this edition of the book because it is particularly relevant to the topic we are discussing. It also conveys the loneliness and agony of a decision maker facing a dilemma in the criminal justice field. Here it is (corrected for spelling and some minor grammatical errors):

> I'm writing this letter more as a point of interest. It might be something that you will want to bring up in class to spawn [sic] discussion. I would only ask that I not be identified directly.
>
> Talking about de-individuation in class and the potential for police/corrections officers to "lose themselves" and compromise their moral autonomy over making moral/ethical decisions, seemed very distant to me. I mean we're not police or correctional officers yet. . . . It doesn't have a direct impact on our lives at the present time. And I know you have told us time and again that the purpose of this course is not to drive certain morals and ethics into our lives but to make us think. It didn't hit home until this week just how much thinking will be involved, and how difficult the thinking process would be in situations where such a problem would arise. It seems now, more than ever, that the process we are undergoing and the concepts that we are learning are applicable to us here and now, and the situations that we face are just a starting point on a journey that will be taken for the rest of our lives.
>
> Before getting to the real dilemma at hand, I will give you a little background about my experience with a policing environment. I have volunteered with [name of police force] for the last nine months. Seven of those months were on a full-time basis (8–4 Monday–Friday, and some weekends). For them I do a variety of

jobs ranging from court work to community-related policing functions. At the beginning of my work with them I did a lot of work with the unit stenographers, typing court documents and preparing legal briefs. I worked alongside two very talented individuals who guided me and provided me with the opportunity to learn many principles of working in a law enforcement environment. One of those individuals, however, was ostracized by the officers and the administration because she didn't really fit the role necessary to work in such an environment. Even that was obvious to me, having worked there for only a few weeks at the time. The officers and the administration would talk about her behind her back, openly criticized her work (that deserved no criticism at all), and generally would go out of their way to make her feel bad some days.

This was apparent to me, but being the new guy I didn't feel it was my place to say anything. About two months into my stay, this individual went off work on stress leave. She would not return to work for five months. Upon returning to work she would last only another three days before having to go off on stress leave again. This individual proceeded to put in 43 complaints against the conduct of the administration, the conduct of officers in general, and even cited misconduct on the part of individual officers in her report.

Now this is where I come in. The complaints have led to a formal investigation on the part of [a senior officer from the head office]. I have received two phone calls and countless e-mails from officers and staff at the detachment warning me of the investigation. I was told numerous times that NO ONE supported this person's claims. And that "I'd better not say anything if contacted," as if to cover up the obvious mistreatment of this employee. The investigator contacted my residence on Tuesday afternoon, however I was unavailable because of exams.

WOW . . . what a thing to think about. Now I know the right thing to do would be to tell the investigator exactly what happened, honestly and truthfully. However, I have become great friends with all of the officers to the point that I feel like an honorary member of their "subculture" if you will. I have been told on numerous occasions by many of the officers that they view me as a colleague and friend, not just a volunteer/student. I really don't want to do anything that would jeopardize that relationship which has taken so long to develop. (Who knows, someday these friends may be my superior officers.) As the dilemma unfolds I can't help but recall what you told us in class, and to search deep inside myself for the answers. They're not so apparent.

I'm not sure what to say to the investigator, however, I do know that whatever I say will be a very difficult decision to make. I am now starting to understand the thought process you were talking about. And I find myself applying the ideas that you taught us in class. I understand the background and the theories behind the material, but I'm just starting to grasp the thought process, and the understanding of myself that will be required in order to make these decisions and work in such an environment for the rest of my life. Little did I think that those situations would creep up so soon in my life.

In the policing examples, I included one incident involving a trainee police officer. The following similar example involves a trainee corrections officer.

An inmate was admitted to a detention center to await trial on charges of having had sexual intercourse with a two-year-old child. One of the officers decided this was such a despicable offense that he was going to exact some measure of revenge for it. He went into the inmate's cell, picked up the Gideon Bible and hit the inmate in the face with it. The inmate's eye was cut, and subsequently swelled shut. Several officers, including the trainee officer, were in the vicinity and witnessed the assault. Later when a lieutenant was investigating the incident, each officer, including the trainee, was asked what he knew of the incident. They all knew what had happened but denied having any knowledge. The trainee later recounted to me that the decision for him was a no-brainer—there was no way he was going to tell the truth because it would have been too costly for him. In all likelihood, his future in corrections would have come to a grinding halt had he broken ranks and told what he saw.

Another of my students, after completing his field practicum at a federal institution, wrote the following. I think it sums up the pressure to conform and adds an air of reality to the predicament officers may find themselves in.

I feel we are at work to do our job properly, so we can pay our bills and provide for our family. If someone else is doing something wrong that isn't any of your business, I think a person should stay out of it. . . . I just feel that a person has just too much on the line to rat someone out and risk losing your job. Whether it is the inmates or the staff, no one likes a rat. If people find out you are a rat nobody will trust you and chances are nobody will want to work with you either. If you go to work every day knowing that nobody likes you then your day at work will not be fun. Before my field placement at [name of institution] I thought I would do the right thing every time even if it cost me my job and my coworker's trust. Now my view has totally changed. I have witnessed the staff subculture and it is very strong. There was an officer at [name of institution] who was a rat and nobody liked him. When he came to the post there would be no conversation and he would usually go to the computer room to be away from everyone because he probably felt very awkward. I don't think I could deal with that for forty hours a week for a lifetime. Even though I may sound heartless, I am being realistic because I want a successful future for my family and me.

The pressure to conform, as one former officer put it, is a very powerful pressure.

These are examples of tough situations for officers, whether they are trainees or veterans. How can we expect officers to be their own persons under such circumstances? Might we be expecting too much of them when we say that we would like them to do the right thing? We shall return to these difficult questions later in the chapter.

At the time of this writing, there is a "joke" making the rounds of the corrections system in my state—one that one of my students told me is related to all new staff and field practicum students. The joke carries with it a message that seems to fit the examples I have given.

Question: How many correctional officers does it take to push a young offender down the stairs?

Answer: None. He fell.

Now it would be inappropriate to suggest that the kind of brutality depicted here in any sense characterizes the behavior of corrections officers or police officers. Perhaps this is an illustration of the kind of dark humor that often marks their conversations. Nonetheless, there is an implied message here—troubling one that it is—that all officers, including trainees, on occasion witness an incident but report something different, in the interests of protecting each other.

Many of us can recall watching the evening news way back on March 3, 1991, and being both amazed and sickened as we watched the videotape of a motorist by the name of Rodney King being beaten by members of the Los Angeles Police Department. But let us imagine for a moment what might have happened had the police officers at that scene been asked to account for their actions before they knew they had been videotaped. Do you suppose their description of the event would have been consistent with what we watched on our television screens? Or do you think they might have reported things very differently in order to protect one another?

These are just a few examples of how our judgment can become obscured, and how the moral choices we make can be seen—at least retrospectively, if not at the time—to be faulty.

With respect to incidents where excessive use of force is the issue, one officer I interviewed asked rhetorically: "Are you holding that guy down a little bit longer to gain approval from your fellow workers? Are you using excessive force for that approval, for that acknowledgment, 'you're all right, you're one of us'?"

These are searching questions but it is important to ask them, for they center on aspects of our value system concerning how we treat other human beings.

Reflections

Would you have charged the brother of the murdered police officer for "mooning"? If yes, why? If no, why not?

"Rookie officers emulate the behavior of field training officers; they pick up very quickly their good habits and the bad ones." What are your views on this statement?

All of the examples related above are examples of officer wrongdoing. Officers found themselves responding in ways that, privately perhaps, they would regret. Why is it, do you think, that occasionally we find ourselves behaving differently when we are with others than we might if we were alone?

Does the fact that you are dealing with an accused person who is alleged to have committed a particularly despicable crime mean that the normal rules of professional conduct can be temporarily suspended?

The Moral Struggle

In some of the examples discussed above, the officers involved actually admitted to knowing that their conduct could probably not be justified. In the other examples, where we do not know what the officers were thinking, would it also be a reasonable assumption to say that in all probability they knew that what they were doing could not be justified?

To think of the purpose of ethics education in the justice system as an activity designed to help officers know what is right or wrong is, in my view, a redundant exercise. There may be some isolated cases in which, either because of moral bankruptcy or the gradual blunting of moral sensibility, an individual has lost sight of what is right and wrong. Generally, however, officers know what the right thing to do is in most given situations. The purpose of ethics training, then, rather than being an attempt to determine what is right and what is wrong, may be an attempt to help officers with their struggle when they experience personal moral conflict either when working alone or in their colleague group. Such a purpose can only be realized, I think, if the ethics educational experience is one characterized by empathy on the part of all concerned—teacher and student alike. If the experience is one in which participants feel they are being judged or manipulated, this would be a prescription for failure—painful to both teacher and student.

The Potential Cost: Moral Conflict Can Make You Sick

Let me offer an idea not generally discussed in any literature that I have encountered. I would suggest that an individual officer's sense of well-being or self-esteem can be positively or adversely affected by how he responds to these moral struggles. I would further suggest that there is a link between how an officer deals with moral conflict and the stress an officer can subsequently experience. Though I can offer no empirical evidence, there is some anecdotal evidence supporting my view, arising out of the interviews I have conducted and from what several ethics seminar participants have confided to me. Additionally, one or two writers have touched obliquely on the subject. I am going to expand on my idea in the hope that the discussion will become a topic of personal thought and class debate for you and your student colleagues.

Ethics education, therefore, can be seen in terms of its usefulness as a stress reliever—as an aid to an officer's sense of well-being. It may also be seen as an aid to an officer reaching retirement relatively unscathed by major career incidents.

It is important that we keep this conversation in perspective. It is not being suggested here that officers encounter these crucial moral dilemmas on a regular basis. These are the more significant moral struggles that may be experienced only occasionally during an entire career. But, as few as they may be, they can be defining moments. How they are responded to can, to a large extent, establish the tenor of the individual officer's professional and personal life. Persistent irritability, relationship breakdown, alcoholism, tarnishing of reputation, loss of colleague respect, or loss of job are just a few examples of what can happen to an officer if found guilty of responding inappropriately to a moral struggle. And any or all of these effects, if experienced, can have a detrimental effect on an officer's general state of health.

We referred earlier to researchers Elizabeth Grossi and Bruce Berg (1991). They wrote a journal article called "Stress and Job Dissatisfaction Among Correctional Officers: An Unexpected Finding." Interestingly, in the context of discussing corrections officer stress, they refer to six stressors, one of which is "that one cannot always act the way either one would choose to, or the way the public might expect them [sic] to." This suggests to me the idea of moral conflict and moral struggle. It is significant that these two writers connect this struggle with the potential for stress. They add: "Again, it may be that in order to attain a satisfactory level of peer support, and reduce conflicting pressures endemic to correctional institutions, correctional officers must compromise their personal values and interests" (p. 80). Such compromise, if experienced, would constitute a fair amount of personal stress for most of us—would it not?

One of my university philosophy teachers, Clive Beck (1991), used to talk about how our value system and our ability to follow it can have an effect on our sense of well-being, or what he called the "good life." If we lose touch with our basic values, he suggests, our enjoyment of life can be impaired. I remember Beck differentiating between *moral values,* which he listed as carefulness, responsibility, courage, self-control, reliability, truthfulness, honesty, fairness, unselfishness, and *social and political values,* such as peace, justice, due process, tolerance, participation, cooperation, sharing, loyalty, solidarity, citizenship (p. 3). You will notice that "loyalty" and "solidarity" are listed as two of the social and political values. And worthy values they are—in the right context. But what if they conflict with other values? What if, as a result of embracing loyalty and solidarity as values above all others, the result is a diminishing of individual courage, or truthfulness, or fairness, or honesty, or justice, or tolerance, or due process? If the latter values matter, and clearly they do for most correctional or police officers, there is going to be conflict and struggle, where one's decisions may well have a direct effect on one's sense of well-being.

My former teacher has a rather unusual way of looking at morality, and makes the point that it is generally unhelpful to blame people or to make them feel guilty for their choices. Rather, he sees a person's immorality as a problem for that individual, something that reduces well-being. He sees such people as unfortunate because they lack the motivation (and courage?) to do what is best for themselves (p. 11). So it would not be too much of a stretch to link the idea of ethics education as it relates to helping us deal with moral conflict with the successful management of stress in our professional lives.

Here is a rather graphic example of how a person, a corrections officer in this case, can have his sense of well-being affected by an act of immorality. The example is from Kelsey Kauffman's book, *Prison Officers and Their World.* The officer is talking to the author after a brutal beating of an inmate by officers. "Maybe I feel what I am saying on his hurt is my hurt. Because I felt so lacking, so missing something that I didn't do something. I was totally ashamed, totally embarrassed that I had to say that I had seen, and I am really, really sorry. And I'm not sorry for him, I'm sorry for me that I didn't have the guts or whatever you want to say, to do something at that time. And it, it eats away afterwards" (pp. 233–234).

Moral Weakness: Moral Courage

One of my ethics course students wrote a comment in her journal that impresses me for its honesty, but at the same time troubles me because it shows poignantly how she has great difficulty dealing with her moral struggles. "I know that I don't always do what is 'right' though I have a sense of what that 'rightness' is," she writes. "It is a constant source of disappointment to me, and I can't seem to rise above it. It's horrible to know what the proper moral stance is and not have the strength to take it. This is another reason for my self-dislike." Clearly, here is an example of a person who is paying a fairly significant personal price for her moral struggle and for what she considers to be her lack of moral courage. She typifies, I think, what Beck means when he says that a person's immorality is a problem for that person.

Rhetorically, this person then asks in her journal a searching question: "Are those who don't see anything wrong with an unethical position happier than those who know the difference, but can't take a stand?" If my former teacher is correct in his thinking, the answer to this question is "yes." It is only at the point where we have a good idea about what we should do, but feel we can't do it, that the serious, health-harming conflict can begin to trouble us. The personal stress begins precisely at that point when we know what we should do, but feel that we do not have the moral courage to do it. Often, I would suggest, our courage is lacking because we fear the consequences of taking a stand. And what I have heard consistently from officers is that what they perceive to be moral failure on their part comes with a price.

One correctional officer trainee referred to this issue of moral courage when he said: "Do you stand alone and become a leper or do you let peer pressure and group think compromise your ethical stand? This is a hard question when you have to feed your family." There is, of course, no easy answer to this question. This person's concern is a legitimate one.

These uncourageous responses, often precipitated by the very real fear of reprisal, can cause many of us (myself included) to respond to the moral struggle by taking the safest route, the route of least resistance. As a result, the decisions we often make are ones that we would prefer not to make if we were stronger. We often take the route that we think will cause us the least amount of grief with our peer group. And often we neglect to think of the private angst that may result from the moral choices we make, and how that may impact on our sense of well-being—on how we feel about ourselves as persons.

Gail Sheehy (1977), in her book *Passages: Predictable Crises of Adult Life,* sees true adult development in the achievement of what she calls "autonomy" and "self-sufficiency" (pp. 52–56).

But this is a difficult enough goal to achieve in our "regular" lives away from the job. It is a much more difficult goal to achieve for police and correctional officers who can be influenced by the subcultural constraints of loyalty to colleagues.

In Chapter 1, reference was made to the Somalia inquiry in Canada. That inquiry into the killing of a Somali youth by some Canadian soldiers on a peace-keeping mission, has become characterized, I think many would agree, by what appears to be stonewalling and a conspiracy of silence on the part of many of the witnesses. In addition, many attempts have been made by the military to discredit the physician who was the first person to raise a concern about the way the youth died. The physician appears to have shown great moral courage, while many other witnesses have not. It is possible the others have not because they fear that they will have to endure ostracization and discrediting similar to that experienced by the physician. But the physician was able to summon the courage to expose what he considered to be a wrongdoing.

What enables some individuals to risk everything to take a moral stand? One writer, Thomas Lickona (1976), a social psychologist, asks a similar question: "What enables a person to maintain moral perspective in the face of strong situational pressures?" (p. 124). I guess if we could answer that question we would all probably be much stronger in the face of moral conflict. One prolific writer in the area of morality, Lawrence Kohlberg (1984), thinks part of the answer may lie with what particular "moral stage" an individual has achieved. Kohlberg talks about six stages ranging from stage one—representing an unthinking, undisciplined moral sense—to stages five and six—representing what Kohlberg calls "principled moral thought" (pp. 454–459).

Thomas Lickona (1976), in an article "What Does Moral Psychology Have to Say?," refers to a U.S. soldier in the Vietnam War. You may remember the My Lai massacre during which an entire village was burned and many of its inhabitants, including women and children, were killed by U.S. soldiers. The soldier's name was Michael Bernhardt, and he was the only ground soldier who refused to participate in the massacre. He was found to be operating at a high stage in Kohlberg's definition of moral reasoning, and here is a statement Lickona attributes to him: "The law is only the law, and many times it is wrong. It's not necessarily just, just because it's the law. My kind of citizen would be guided by his own laws. These would be more strict in a lot of cases, than the actual laws. People must be guided by their own standards, by their self-discipline" (p. 125).

You will notice there is no sense of blind allegiance in this statement. When one considers what the pressure must have been like for Bernhardt to conform with the rest of his highly trained soldier colleagues, it is amazing that he was able to remain his own person.

In late 1998 I attended a police ethics conference in the United States. One of the guest speakers was Mr. Hugh Thompson, a retired U.S. Army captain. Captain Thompson had also been involved in the My Lai affair as a helicopter pilot. For many years after the war he had been castigated as a traitor. His offense was that when American soldiers were shooting unarmed villagers he landed his helicopter between the soldiers and the villagers and ordered the soldiers to stop shooting. He went further. He announced over a loudspeaker that if they did not stop shooting he would order his helicopter gun crew to open fire on them.

Many years later, Captain Thompson is being hailed as a hero. The documentary program *Sixty Minutes* did a story on him and took him back with reporter Mike Wallace to Vietnam where he met with some of the villagers he had helped save. And though they waited 30 years to do so, the Pentagon has also recognized him as a hero for the role he played.

Captain Thompson, in his speech at the conference, appeared to be a very self-effacing individual. When he was asked why he did what he did, he responded: "I don't see how I could not have done what I did."

Both Michael Bernhardt and Captain Thompson showed a rare courage. They somehow knew what it was they should do, and did so without fearing the consequences.

Of course, you may ask what kind of army would one have if all soldiers danced to their own drummer! What would happen to the *esprit de corps* so essential to a group of soldiers fighting together?

Keep in mind here that we are discussing situations *in extremis,* as it were; situations that cause a significant crisis of conscience in an individual. Generally, you can still have your togetherness and solidarity and camaraderie—but only up to the point beyond which an individual would betray her sense of morality.

I remember reading a newspaper article in Australia entitled "Why Loyalty Is a Two-Way Street." One question posed by writer Adele Horin was: "Is a loyal Nazi better than a disloyal one?" She made the point that loyalty "can blind you to higher loyalties, to truth-telling, justice, the common good." After pointing out that loyalty to a group can blind us to what she calls "wider commitments and obligations," she adds: "For example, the Blue Wall of Silence around police is a wall of loyalty and fear that keeps buddies true to each other at the expense, sometimes, of the good name of the force" (*The Sydney Morning Herald,* July 9, 1994). It is significant that she uses the word "fear" here. The fear, presumably, would be that of being ostracized by the group.

One of my current students, as I write this chapter, has a brother on a Canadian police force. His brother has told him that the worst thing that can happen to a police officer in his professional life is to be ostracized by his colleagues. It would demand considerable courage to take the kind of stand Michael Bernhardt and Captain Hugh Thompson took.

But let's return to this idea that *not* being courageous can sometimes, in a real sense, have an equally devastating effect on us and our careers. William Tafoya (1995), in an article called "Ethics and the Realities of Life: Surviving the Vortex," graphically depicts how one police officer got caught up in what Tafoya calls a "vortex": a kind of whirlpool that swallows or consumes everyone caught up in it. The incident he refers to is one in which a group of police officers in a major metropolitan police department were terminated for their part in the beating death of a motorist and the subsequent cover-up. One of the terminated officers spoke with Tafoya about the event several years later, and much of the article is a verbatim account of the interview. The article is a gripping, and sobering, account of what can happen to an otherwise good, but uncourageous person. In his introduction, Tafoya expresses something amounting to incredulity that an otherwise good and intelligent officer would feel compelled to lie. This was the case despite the fact that the officer had come from a solid upbringing in which values and doing the right thing had been highly prized, and the fact that he knew what he was doing was wrong. Tafoya points out that even though this officer felt the cover-up was going to fail, he "nonetheless believed he must go along with the attempt to do so" (p. 576).

We clearly do not have the space here to include the long transcript (which in my view ought to be mandatory reading for every police officer), but I do think it important to include some of what the former officer said to Tafoya. In the transcript of the interview, Tafoya notes in several places where the former officer either began to, or actually lost, his composure. Keep in mind that the incident had taken place several years earlier, and that he had made a relative success of his life after policing. Despite this, it is clear that he had suffered severe emotional and psychological pain at the time of the incident and in the ensuing years. Here are some of the things the former officer said:

> Yes. [in response to a question about the potential for retaliation from colleagues] This goes back a long way, before I joined the police department. I was socialized early and often in the police culture by friends who were police officers. From them I heard repeatedly that the most important lesson to be learned was sticking together. One friend [Fred], suggested that if I ever saw anything that I thought was improper—and was a rookie—I should keep quiet, go back to the car, and make like I was writing a report; stuff like that. . . . A woman in our training academy class had previously done an internship with the department's Detective Bureau. . . . We asked her about what it meant [the expression "Hand-up guy"]. She said rather sarcastically that "Hand-up" is the worst insult one police officer can give another. A "Stand-up" cop is one who covers for a fellow officer. Everybody likes such officers, trusts them. Such officers go to bat for one another. But a "Hand-up" cop is likely to tell the truth and not cover for a fellow officer. It was then that I first began to understand that in the police culture, telling the truth was not always a virtue.

You will have noticed that these sentiments in a real way echo much of what we have been discussing in the book to this point. Now let's take a brief look at some of the things this fellow had to say about the effects on him personally of his inability to take a moral stand.

> I grew up in a little Christian Reformed community. I'm a religious person. I know what the truth is. I know you are supposed to be honest. That is probably what has bothered me more than anything else. I know what I did was wrong. I knew it was wrong when I did it. I was sick with myself; still am. . . . To this day I struggle with this. . . . Oh God (becomes very emotional). What I did, I'm still ashamed of what I did. A comment a good friend made still haunts me. He said, "Integrity is like virginity. Once you lose it, it's gone." (begins to cry). It's been a lot of years; that's an awful long time. I don't think as though I've got closure yet. I don't think I am going to wake up and find out it was all a bad dream. It happened and I'm still having problems with it. (pp. 577–580)

In my view, this is a graphic and disturbing account of how an otherwise principled officer, in the interests of being seen by his colleagues to be a "stand-up" officer, was prepared to put at risk, if not lose, everything he valued in life, including his reputation and career. For many of us, as for this former officer, such an experience would certainly result in a health-harming loss of self-respect. Indeed, at one point in the interview the former officer said that the primary reason for his agreeing to the interview was to try to come to terms with the whole issue. "That's why I'm letting you do this to me [the interview]," he said. "This is a catharsis for me.

I've gotta stop somebody else, some other police officer, from doing this. I don't want somebody else to go through what I did" (p. 583).

Of course, one big question that has yet to be addressed is "What if we make a decision that to be ostracized by colleagues would be more stressful than anything else that can happen to us?" This is a question that needs to be asked, and one we will address in Chapter 8. The former police officer in Tafoya's interview was asked what police departments could do to help officers like him who essentially open themselves up to becoming corrupted by responding in inappropriate ways because of subcultural constraints. One disturbing expression he used to describe his dilemma was "I guess I was damned if I did, and I was damned if I didn't" (p. 580). This important issue will also be addressed in Chapter 8.

One final thought on the discussion in this chapter specifically and in the book generally up to this point. There have been times during the process of writing this book that I have felt compelled to complete it because of certain "events" that occurred during its writing. One such event occurred two days ago. I had just returned from a two-week trip overseas and there were several messages waiting for me on my voice-mail at the college. One message was from a former student who is now working in corrections.

> I am phoning to apologize to you, John, because when you taught us the ethics course I did not accept what you said to us that there may be moments in our career when we would find it very difficult to make the right decision. I always felt that I would never experience such difficulty because I was convinced that I would always be able to do what I thought was the right thing, that there would be no question about it. Well, I'm phoning to tell you I was wrong, and I want to apologize to you for not believing you. I am now confronted by a situation at work where, if I do what I think I should do, I will lose my job. So I just wanted to apologize to you and to say you were right in what you said in class.

Now it doesn't matter to me whether what I said in class was "right" or not. I was simply attempting to get the students to think about issues of professionalism and moral responsibility. What I did find upsetting was the tone of voice of the student. She did not leave a phone number and I do not know where she is working, so I have been unable to return her call. She is clearly struggling with the problem in a way, it seems to me, that is adversely affecting her life.

The personal price we can pay in attempting to deal with our moral struggles can be a significant one. And this is where, in my view, ethics education approached in an empathetic way may be of some help.

Reflections

Would you agree, for the most part, that officers know what they should do in most situations, and that, if ethics education is geared to teaching what is right and wrong, it might be a redundant activity?

What is your view of the idea that our general sense of well-being as individuals is connected to how we resolve our moral struggles? What is your experience?

Have you ever considered how you would deal with a situation in which loyalty to colleagues may conflict with the values of justice, fairness, or honesty?

Why is it, do you think, that individuals such as Michael Bernhardt and Hugh Thompson are able to take a moral stand even when the consequences are likely to be painful?

Is a loyal Nazi better than a disloyal one?

A friend of the former police officer interviewed by William Tafoya said: "Integrity is like virginity. Once you lose it, it's gone." How do you react to this statement?

Summary

In this chapter we have suggested that the solidarity norm in officer subculture constitutes fertile ground for officer wrongdoing. We discussed some actual incidents both from corrections and policing to illustrate how this may be the case.

In the discussion it was suggested that officers, for the most part, know what is right and wrong, and that ethics education that is geared toward helping individuals know what is right or wrong is probably redundant. It was suggested that the purpose of ethics education should be to help officers cope with their moral struggles. As a result, on occasion officers may be able to avoid falling into a similar trap—or, to use Tafoya's term, "vortex"—to that in which the former police officer interviewed by Tafoya found himself.

One of the benefits of ethics education, then, is that directly or indirectly it can contribute to an officer's sense of well-being. That is to say, if an officer, as a result of an ethics education experience, manages to deal with her moral struggles in a personally satisfying way, her sense of self-esteem would be kept intact—if not enhanced. Of course, the converse would be equally true, as was so dramatically illustrated in the case of Tafoya's interview subject.

8 Tools for Moral Decision Making

I have always had the notion that we never truly think while in college. All we do is memorize.

—Journal entry, justice student in ethics course

The answer to a particular issue or problem isn't the important thing, the important thing is how we arrive there.

—Journal entry, justice student in ethics course

The first roadblock I encountered on my ethical journey was leaving my emotions and feelings at the door. . . . I slowly realized how cloudy an issue becomes when you include your emotions and feelings.

—Journal entry, justice student in ethics course

To me ethics is what we choose as right or wrong, how it compares to what society decides is right or wrong, how we choose to react to the differences, and finally, what we choose to do when we are confronted with the conflict of what we should do.

—Journal entry, justice student in ethics course

I feel that it is very important to look at things critically. Far too often people take things at face value, for example, as in information presented in newspapers or on TV. I think we need to consider where the information is coming from, who is saying it, and what is really being said. By thinking critically about things I feel we can become more educated—and safe.

—Journal entry, justice student in ethics course

Chapter Objectives

After reading this chapter you should be able to

- *Explain why dilemmas are usually complex.*
- *Explain why it is that however we respond to some dilemmas, we will feel that we are in a "no-win" situation.*
- *Explain what is meant by the phrase "the choice really may be one between 'self-respect' and 'colleague respect.' "*
- *Identify the two ideas that some officers have expressed that may help to explain how some officers deal with dilemmas they experience on the job.*
- *Identify the "tools" that may help our decision making when we are faced with a moral dilemma.*
- *Explain why there may be a gender difference with respect to individual officers "setting limits."*
- *Explain what is meant by* ***setting limits.***
- *Explain the principles of* ***benefit maximization*** *and* ***equal respect.***
- *Explain what we mean by the* ***bell,*** *the* ***book,*** *and the* ***candle.***
- *Explain what we mean by the* ***imaginary video camera.***
- *Explain the general idea of* ***critical thinking.***
- *Explain what we mean by* ***critical*** *in critical thinking.*
- *Identify five elements of critical thinking.*
- *Explain how thinking about moral stages may help our moral decision making.*

Introduction

By way of introducing the discussion on the tools of moral decision making, consider two related ideas raised in the preceding chapter, one in the form of a question and the other in the form of a statement.

The question is this: "What if we decide that to be ostracized by colleagues would be more stressful than anything else that can happen to us?"

The statement is as follows: "I guess I was damned if I did, and I was damned if I didn't." It was expressed by the subject of William Tafoya's interview when describing the dilemma that led to his wrongdoing and subsequent dismissal as a police officer.

The intrinsic nature of any dilemma is that it would normally beg no simple solution. If the issue confronting us is truly a dilemma then making a decision about it is probably going to be difficult. Sometimes even choosing to do what we consider to be the right thing does not necessarily mean that we are going to be completely at peace with our decision.

This would particularly be the case if we elected to make a decision that could implicate a colleague officer in some way. I think this is the kind of dilemma with which Tafoya's former police officer subject found himself confronted. If he had made a decision not to get caught up in the lying and fabrication, then in all likelihood he would have had to deal with the subsequent wrath and rejection of his colleagues. He would probably not be trusted by them again, and would have to endure the anguish of being labeled a "hand-up cop," and all that can go with that kind of labeling. On the other hand, having made a decision to stay solid, he experienced not only the public indignity of being fired as a crooked police officer, but also for many years has had to struggle, day in and day out, with his own personal demons arising from the incident.

No-Win Situations

These kinds of situations often appear to be, in a very real sense, no-win situations. On the one hand, you can choose to stay solid, to be seen as a "stand-up cop," and continue to earn the acceptance, loyalty, and support of your colleagues; on the other hand, you can choose to act according to the dictates of your conscience if it is leading you in another direction, and run the distinct risk of losing the acceptance, loyalty, and support of your colleagues.

It would seem that the difference between the two choices is that you can opt either to continue to be admired and respected by your colleagues but lose some or all of your self-respect, *or* you can opt to hold on to your sense of self-respect but lose the respect of your colleagues.

Of course, there may be many occasions when the retention of both self-respect and the respect of your colleagues would be the result of your decision. I am not suggesting that all hard choices necessarily result in the "either this . . . or that" scenario. But there will almost certainly be times in your career as an officer when you will be confronted with the kind of dilemma we have been discussing here—with a choice that really is one between self-respect and colleague respect. And when these situations arise, they can, as we have discussed, be extremely difficult to resolve, and result in personal anguish.

It is at these moments that one's sense of self, and who one is as a person, is put squarely on the line.

Two Ideas

Perhaps it is appropriate to mention here, as an aside, that I have had some interesting conversations with officers about this topic. They have expressed two ideas that will perhaps provide interesting material for your personal thought or for discussion with your colleagues. Perhaps some of you may think these ideas worthy of exploration, and I would encourage you to do some further research on them.

We don't have time in the context of this book to discuss the ideas in detail, nor am I going to offer any comment on them, but here they are:

First, a few officers have suggested to me that some officers are able to cope with dilemmas that have the potential to put them at odds with their colleagues, simply by regarding loyalty to and solidarity with colleagues as the *preeminent* value in their professional lives. In other words, they have decided that to abide by this principle is the ultimate morality. This one-dimensional thinking, of course, can enable individuals to get out from under some moral struggles that are essentially multidimensional in nature and which do not beg a simplistic solution. It may be an effective way of avoiding anguish for some officers. Whether or not this approach would count as an intellectually honest one is a moot point.

Second, some officers have suggested that when they are at work, because of the unusual nature of their job and the character of the socially deviant "clients" with whom they normally work, they can mentally and emotionally justify *suspending* the morality they operate by in their private lives. In other words, they can rationalize allowing themselves to be governed by a different, and somewhat questionable, morality on the job. (This sentiment has been expressed almost exclusively by male officers. Perhaps you can remember our discussion in Chapter 6 about this topic). If, indeed, this rationalizing does happen, it would seem here that one's moral sense is compartmentalized—that one's "real" moral sense is practiced and adhered to in one's private life, where one remains committed to truth-telling and fairness, and so on; and that the other moral(?) sense in the workplace allows one to engage in what I call "means-end" morality, where the normal tenets of morality, such as the ones we have mentioned, can be suspended or ignored. I trust you will have some fun debating and discussing these ideas.

Reflections

What is your view of the idea that to be ostracized by colleague officers for taking a moral stand might be one of the worst things that can happen to an officer?

If you were confronted by what you thought was a "no-win" situation, what factors would probably influence your decision making? Attempt to "anchor" your response to a specific situation.

Is it reasonable to talk about "self-respect" and "colleague respect"? Or, does this way of thinking simply not recognize that, for the officer, the two are invariably linked?

What are your views on the two ideas expressed by officers as a way of explaining how some officers are able to deal with moral conflict?

Let's return to the topic under discussion. We have suggested how extremely difficult it can be to be confronted with a no-win situation. We are now going to discuss some "tools" we can use in our moral decision making. We should come at those decisions understanding that there is seldom a perfect solution, and that there is probably going to be some pain whatever we choose to do in a given situation. The tools, hopefully, will enable us to make decisions that give us the greatest degree of comfort.

I guess one test of a decision we are wrestling with might be to reflect on it at three in the morning on a night of fitful sleep, and see how we feel about it!

If by using one or more of these tools you are helped to make the best possible decision for you and for others affected by that decision, then the time and effort you commit to considering them may be more than justified—at least in terms of your peace of mind and sense of well-being. More important, perhaps, thinking in this way may help at some point to save your career.

Now while it is true that sometimes in our moral decision making we must, as it were, make the best of a bad job, that is, make the best decision we can under the circumstances, be very much aware that you may experience times, especially if you apply the following tools to your decision making, when you may come to a pretty certain and compelling decision about what is the "right" or ethical thing to do in a given situation. If you act on such a decision because you are sure that this is what you should do, you may still experience pain and anguish because of the way others may respond to your choice. Taking the moral high road can be costly even when we are convinced this is the route we should follow.

Think about these tools. Apply them in your daily life. You may find them useful. And you may find they will help you avoid a jackpot (in the negative sense) someday!

Setting Limits

This is a technique that some officers have told me helps protect them from becoming caught up in potentially compromising situations. They simply indicate to colleagues, in a pleasant and courteous way, what their personal limits are. They may do this regarding, for example, using too much force, or gossiping, or sleeping on night shifts, or accepting gratuities, or whatever the case may be. These questionable practices are simply cited as examples, and are not meant to imply that any one of them or a combination of them normally characterizes the behavior of officers.

Setting these limits and making them known to colleagues can, I am told, cause a colleague who might be contemplating responding unethically to a given situation to have second thoughts about doing so in your presence.

An example might be a situation in which an officer partner on night shift tells you he wants you to "spot each other off" as you alternately take turns sleeping. You might say to him, by way of establishing your own limits, "I don't feel safe when either one of us is sleeping, so I would prefer it if we both stayed awake." If he tells you to take a running jump, so to speak, you may then say, for example, "I will not sleep, and I would prefer it if you did not. It is not safe. But if you continue to sleep I may be left with no other option but to take it further. I really do not want any unpleasantness between us, but the choice is yours." To set our own personal limits in this way can be difficult for us, but it can save an awful lot of grief later. What do you think?

Perhaps it should be added here that many of the officers who have advocated this technique have been women. My research shows clearly that female officers more than male officers use this technique effectively. We can hypothesize why this may be the case, and I hope that you will give some thought to the issue. The research also shows that when male officers are able to use this technique without running the risk of alienating their colleagues, they generally enjoy the comfort of being viewed by their colleagues as credible, experienced, and reliable in an emergency. It could be problematic for a rookie officer to use this technique, especially if she is perceived by more experienced colleagues to be superior to them in some way. A new officer would have to set personal limits cautiously and diplomatically, otherwise he could be setting himself up for an interesting orientation to the job! This is not necessarily to suggest, however, that if you are a trainee officer you should not consider the advantages of adopting this technique.

Reflections

Is it realistic, do you think, for a new or relatively new officer to "set limits" with colleagues?

If you answered yes, what are the risks, if any?

If you answered no, try to explain why you feel this way.

Some male officers have expressed the view that they perceive themselves to be much more subject to the potential for what they call "street justice" than are female officers. What do you think? What is your experience?

What is your view of the idea that experienced, proven officers are more able to set personal limits?

Ethical Principles

There are some writers who think, at those times when there is time to think (and I realize that this may not always be the case in corrections or policing), that it can be helpful to incorporate the consideration of certain ethical principles into our pondering a dilemma. Three university professors—Kenneth Strike, Jonas Soltis, and Emil Haller (1988)—have written a book called *The Ethics of School Administration.* In the book they explain skillfully and demonstrate how incorporating some thought about ethical principles into our moral decision making can help us arrive at more informed and probably more correct decisions. In the context of this book we do not have the time, in any kind of depth, to discuss the practice, so I would recommend that you read the book if what follows is of special interest to you. The practice is what Strike et al. call "the application of principles to cases" (p. 4).

Let me caution you, however, before we discuss the practice of applying ethical principles to resolving moral dilemmas. The ethical principles themselves can sometimes give us conflicting messages about what we should do in a given situation. A second caution would be that incorporating a consideration of the principles into our decision making could potentially rationalize a decision with which we basically feel uncomfortable.

I would encourage you to practice using the principles in your own moral decision making. One way to get used to incorporating them into your thinking would be to use them as you discuss the dilemmas in Chapter 9, and any other dilemmas arising from your own experience that you feel comfortable bringing to class for examination. There are two basic principles:

The Principle of Benefit Maximization

This principle holds that when we are faced with a conflict about what we ought to do about any given moral dilemma, we should attempt, prior to making any decision, to weigh the consequences of our decision. If we judge that one particular decision, among all the ones we could make, is likely to lead to the greatest amount of good for the greatest number of people, then this decision is the one, according to this principle, that we should make. The morality of our action, therefore, is judged against the probable consequences of that action.

This principle can be useful in helping us think through our moral dilemmas. However, of the two principles we are going to discuss, this is the one that is most problematic. We suggested earlier that sometimes these principles can give us conflicting messages about what we should do in a given situation. The principle of benefit maximization is the principle more likely to conflict with the other. This is a principle that can, unless we are careful, be used for the purpose of rationalizing decisions and behaviors about which we do not feel at peace or, more plainly put, which we suspect are wrong.

For example, a couple of police officers may apprehend a suspect who they have a strong instinct (but no proof) is at the root of a string of crimes that have been committed recently. The officers know from their experience that this particular suspect is a bad actor and that the community generally would be safer with him behind bars. You can see already that in a case like this it would be easy to use this principle of benefit maximization to justify, for example, planting drugs on this suspect, in the absence of the officers' being able to come up with any concrete evidence. In that way, they may convince themselves they have served the community well by making it a safer place. In other words a little bit of injustice has led, they convince themselves, to the greater good. Dubious means have led to a good end, so the wrongdoing is justified as well-intentioned and productive.

In the corrections context, correctional officers may feel that to beat a convicted child molester whom they feel has "got away" with a light sentence may cause that offender to think twice about reoffending, again thereby rendering the community a safer place for children.

As reprehensible as child molestation is, we should think long and hard about whether, as peace officers, we have the right to suspend the normal rules of justice, law, and order, in the course of our duties.

The kind of thinking and behavior in the two hypothetical situations above would be difficult to defend, either morally or legally. Many would consider the officers' responses to be illegal and unprofessional. Others would consider them to be unethical, because the actions of the police officers and the correctional officers are the sorts of actions that would normally lead to cover-ups and dishonesty in the interests of self-protection.

Their actions may essentially represent a kind of vigilante justice; the thinking that governs these kinds of actions may not be much different from the thinking that governed lynch mobs.

For the last decade or so this utilitarian approach (another name given to this kind of thinking) to making moral decisions has been embraced to a great extent in the Western world, particularly in North America. But because of the tendency of individuals to use this sort of thinking for the purpose of rationalizing all kinds of questionable behaviors, it is now, in the early 2000s, being questioned as a useful aid.

My view is that although it is a useful concept to keep in mind, the principle of benefit maximization is more likely to help us in an objective way if used in conjunction with the second principle we are going to discuss. The two principles looked at together may serve us well as they provide checks and balances against any self-serving motives that may otherwise be lurking in our decision making.

The Principle of Equal Respect

When we use this principle, rather than concentrating our moral decision making on the potential benefit of that decision to the greatest number of individuals, we concentrate much more on the appropriateness of considering all individuals as being worthy of equal treatment, simply because they are human. The humanity of individuals is the common denominator here rather than, for example, their cultural or socioeconomic status.

You can readily see that when we consider the two hypothetical situations we discussed in The Principle of Benefit Maximization section, and incorporate into our thinking the principle of equal respect, it would be difficult to continue to feel justified in taking the law into our own hands. Would we wish similar treatment to be handed out to our son, mother, or grandparent? How would we feel if we were on the receiving end of such treatment? Would we wish people generally to be treated in this way?

I think we could make a reasonable assumption that most officers would respond to these questions in a way that would indicate they are committed to fairness and justice. See if you think the principle of equal respect comes to bear on the following example.

Among all the wonderful neighborhoods that make New York City such a delightful place to visit, there is an area known as Greenwich Village. If challenged to explain the attraction of "The Village," one would most often describe a kind of world crossroads, where a delicious mosaic of peoples comes together to sample an eclectic representation of art, music, and poetry. Well, you get the picture. Toward the end of 1998, though, full-time Village residents had apparently had enough of the antics of that "delicious mosaic," and they requested relief from the New York City Police Department. "Enough is enough," they said. "We want an end to the noisy bars, the drunk driving, the drugs, the street minstrels. . . . We want things brought under control."

Responding to the call from the community, the New York City Police turned out in force. They even created a name for their task force, calling it "Operation Civil Village." It did not take very long after the start of that initiative before residents were absolutely ecstatic about the positive changes. Seemingly overnight, the bars had turned down the volume on their sound systems, the drug dealers had moved away, the street minstrels were nowhere to be found, and the drunk driving problem seemed to have disappeared. An amazing transformation, and one that met with a broad outpouring of support from the formerly aggrieved residents.

Gradually, though, there was an interesting shift in the level of resident support for the efforts of "Operation Civil Village." After a year or so, public comment began to run along the lines of "Yes, they have done very good things. But I live here. And I'm getting weary of having to produce my identification every time I walk down the street. And I don't think I should have to be stopped at three DUI road checks every time I drive into the neighborhood." In other words, as long as the policing effort is directed at keeping other people in line, I am in support of it. But when that same effort puts me under scrutiny or treats me like those "outsiders" . . . well, I find that uncomfortable, and it must be stopped.

As you might agree by stopping to consider this principle before we make a decision, our thinking about a particular course of action and our subsequent behavior may be radically different than if we had not stopped and thought in this way. And, equally important, the danger can be reduced of falling into the potential trap of rationalizing an act that could get us into deep trouble.

If we think about it for a moment, it becomes clear that the notion of due process has its roots in this principle of equal respect, as does our thinking about the basic rights of individuals in the kind of society in which we live. The concept of due process, though a legal term, is very much a moral term also. As you know, the concept of due process has to do with developing and applying fair procedures for making important decisions affecting the lives of others. There is something ethical about such an approach. As a private citizen I take comfort from it, even though from time to time the process can let us down.

Reflections

Why should we be aware of the inherent dangers of using only one ethical principle to justify a decision?

Think of a moral dilemma you have encountered. How would you use either or both of the ethical principles to assist you in your decision making?

Is an accused person or an offender entitled to benefit from the principle of equal respect?

Is it possible, do you think, to make a good case for sometimes suspending due process?

The Bell, the Book, the Candle

These are tools for moral decision making that have been suggested by Michael Josephson, founder and President of The Josephson Institute for Ethics in California. They are simple and straightforward to use, but can be a most useful check against which to measure a course of action we are considering. I would suggest here—and this is a point that you may wish to think about for yourself and also to debate in class—that if we are contemplating doing something that we feel reluctant to subject to the scrutiny of tools like these, then we probably know already whether we should take that step. Our reluctance probably indicates what we know either instinctively or intellectually that what we are proposing to do is not morally justifiable.

The *bell* simply refers to a question we might ask ourselves when we are about to make a choice: "Are any warning bells going off?" If so, then we should at least stop and reconsider what we are proposing to do. Of course, the warning bells may be going off simply because of a particular way in which we have been socialized or conditioned, where the proposed course of action is still a defensible and ethical response. Normally, however, if the warning bells go off, and we wish to avoid for ourselves much subsequent grief, we would be wise to heed their warning and respond appropriately.

The *book* is a tool that simply refers to laws, regulations, codes of ethics, and other forms of written guidance. So when I am contemplating a course of action there are a number of questions I can ask myself. Is what I am proposing to do lawful? Does it fall within the department's policies and procedures? Does it potentially violate any section of the department's Code of Ethics?

The *candle* simply refers to how the decision might look if it were exposed to the light of day, as it were, or to public scrutiny. If I choose to act in this way and it is examined in the public arena, can I defend my action? Will I continue to feel good about my action, and will I continue to feel justified in doing what I did?

Reflections

Is it possible to quickly review the bell, the book, and the candle in heat-of-the-moment decisions? Or are these tools only useful when considering moral conflicts, where making a quick decision is unlikely to be an issue?

Have you ever consciously used the bell, the book, and the candle, or something approximating them, in your own moral decision making?

If your answer is yes, was the exercise helpful as a guide in your decision making?

The Imaginary Video Camera

We live in a technological age and one of the curses or blessings of this age—depending on one's perspective—is the video camera. Many unsuspecting souls, as we have seen on the news and in documentaries, have discovered to their great cost that they are the unwitting "stars" of video footage. And included among these unsuspecting souls, as many of you will have seen, have been a number of law enforcement personnel.

I am, of course, talking about those embarrassing moments when inappropriate conduct is captured by some amateur sleuth in the vicinity, armed with her video camera. For example, many of you reading this will remember the graphic video footage of the Rodney King beating by police officers in Los Angeles.

On another occasion, on the news the other evening there was again graphic film footage of a police officer, having stopped an unarmed motorist for a traffic violation, inexplicably firing several shots at the driver who later died of his injuries. The driver was unarmed, and had stopped his car, at which point the officer appeared to lose his composure. Without the benefit of the video recording, most of us would have believed the officer's subsequent story that he had been attacked first.

Some interesting events unfolded in New South Wales, Australia, in 1996 and 1997. A royal commission was appointed to look into charges of widespread police corruption across the state. As part of the investigation, some police cars were bugged. Microscopic cameras were strategically placed in the dashboard of the cars. As a result, many illegal activities by police officers were recorded and later produced as evidence to the inquiry. One particularly unscrupulous and corrupt police sergeant, after telling a television news crew that he knew nothing of police corruption, was subsequently shown video footage of himself sitting in the driver's seat of his police car, counting out money he had just been paid by a drug dealer. The dealer was sitting in the passenger seat of the police car, as bold as brass, almost as though it were his office!

You will have heard the humorous expression "You know it's going to be a bad day when a reporter from *60 Minutes* taps at your door!" The police sergeant's experience of watching himself as his corruption was exposed for all Australians to see must have given new meaning to the expression.

These are just a few examples of how we can get caught out by the wonders of modern technology. It would be fair to say that you and I may also at some point find ourselves on the receiving end—having our movements and behaviors captured by a video camera without our permission. And that thought alone can be a useful deterrent to our behaving unprofessionally.

What I am suggesting here is that on those occasions when you or I are beginning to feel uncomfortable, either with what we are thinking of doing, or indeed are already doing, imagining that our behaviors are being recorded may help us to gain a clearer perspective. Thinking in this way may be a quick but useful check at that moment on whether or not our judgment, or our behavior, or the action we are contemplating, is sound and defensible.

As you can readily see, this technique is an extension of the candle technique, in which we ask ourselves whether our actions can be justified in the light of day.

Reflections

Is there a danger that in suggesting the imaginary video camera as a tool, I am playing on the emotion of guilt?

If you answered yes, is there a possibility that this could be construed as a manipulative move on my part?

If you think it might be manipulative, is the manipulation justified if it results in even one career being saved?

If you answered yes, why? If you answered no, why not?

(*Note:* You can see from these questions that this is an issue with which I wrestled prior to my making a decision to include the imaginary video camera as a tool.)

Critical Thinking

Critical thinking is not so much a specific tool to help us with our moral decision making, as a general approach to problem solving that can also help us as we struggle with our moral dilemmas.

You will no doubt be aware, through your own reading and through some of the work you have done in your classes, of the concept called *critical thinking.* There are many books on the subject and I would encourage you to do some research and reading on the topic for yourself. In the context of this text we only have time to touch on the skill.

You will notice that I use the word "skill" to describe the process of critical thinking. It is a skill that we can acquire through learning and discipline, but something that few of us come by naturally.

Briefly, then, what do we mean by the term "critical thinking"? First of all, the skill has little or nothing to do with being critical in the way we normally think of being critical. It doesn't mean we should go around criticizing. Rather, critical thinking means that we should refrain from taking any idea or point of view at face value; that we should examine it carefully before making a personal judgment about it. It means also that we should probably think twice about allowing others to do our thinking for us. It means that we should aim to think for ourselves, to weigh the relative merits and demerits of any idea, to be prepared to look at all sides of a given issue, and to come to our own informed opinion.

The task, however, is not an easy one. Our thinking in general, and the way we look at most issues specifically, is unclear at best, leading us often either to simply accept a point of view without examining it for ourselves, or to cling to our own—often blindly accepted or adopted—position on a given issue.

In the literature, you will find many variations on the theme of critical thinking. Put simply, critical thinking can be summed up in the following way. If we are critical thinkers,

1. we will resist the temptation to jump to conclusions about issues;
2. we will be prepared and able to think independently—that is, to think for ourselves;
3. we will have the courage always to be prepared to subject our own views and ideas to scrutiny by ourselves and others;
4. we will refrain from accepting blindly any point of view or idea that is offered, even (or especially!) if it is offered by someone we consider to be an expert; and
5. we can demonstrate skills of reasoning that would include the ability to honestly weigh the evidence, to trust only reliable sources of information, to disregard hearsay or unsubstantiated opinions, and to think logically about an issue.

You can see that if I discipline myself to think in this way, my response to situations involving rumor or gossip or peer pressure, for example, is probably going to be different from my response if I did not.

I use a handout in my class that my students find very helpful. It is a short piece titled "On the Art of Thinking" by a man called Roscoe Drummond. Unfortunately I cannot cite the article because I am not sure how I came across it. Drummond makes the point that young men and women should be stimulated to think about knowledge, not just to acquire it.

You will notice that at the beginning of this chapter I include a journal quote from one of my students. He or she wrote, "I have always had the notion that we never truly think while in college. All we do is memorize." Drummond said something similar: "I was so busy learning facts and memorizing them so that I could pass examinations that I had no time to think about them." He then goes on to say something that should be taken quite seriously by all teachers including me. "I cannot recall after graduation I ever had the need or the opportunity to use a single fact I had acquired in getting what passed for an education."

In the article Drummond continues to offer some very useful tips for those who wish to learn how to think critically. Here they are:

1. Before you assume to reach a judgment on a given issue, insist on getting all the pertinent facts, not just some of them, which bear on it.
2. Read at least one newspaper and one magazine whose views are likely to run counter to your own. We need information we may not always relish and we need to test our insights and our opinions against differing information and differing opinions.
3. One way to help get a right answer is to ask the right questions. They constitute the beginnings of what can lead to a sound conclusion.

4. One cannot hardly deem himself adequately qualified to defend his own view on a given matter unless he has equipped himself to expound the contrary view. It's a healthy growing experience. (My students are always asked to take the view opposite to their own in the class debates we hold. They find the exercise a difficult one but find that it stretches their thinking and forces them to look at other points of view. Try it sometime—you might like it!).
5. All truth is rarely on one side in any discussion of temporal affairs. Sometimes the balance of fact and good sense will be fairly even. But when choices have to be made, this need not affect the firmness of your conviction.
6. Don't hesitate to change your mind; it's a rewarding experience. Don't hesitate to hold to your convictions when you are persuaded you are right.
7. A yearly mental housecleaning can be helpful. Remove from your mind all your convictions and prejudices, examine them honestly and then decide which should be put back and which, if any, should be discarded.

These tips are good ones and worth keeping in mind when we wrestle with difficult issues.

Let me share something with you that I tell my students. Keep in mind that it is a personal opinion of mine and I am more than willing for you to hold it up to scrutiny, to see whether there is any merit in it. I tell my students how I believe we often develop our value systems. I explain to them that on occasion we accept ideas and values from others—normally significant others, for example, parents, teachers, sports coaches, or our church or school leaders—and we put them on as we would an item of clothing. Just as we generally only put on clothing we like, we tend to accept the values we "like." The problem is that if indeed there is any truth in what I am saying, many of our values cannot be said to be truly our own. We have adopted them, as it were, without holding them up to careful examination.

In essence, it is something like building our values foundation on sand. Because they are not truly our own values, they can be readily shifted, or challenged or changed. Because these adopted values are not our own, our investment in them can often be shallow, causing them to become quite vulnerable to any wind of change.

Reflections

Would you agree that the ability to think critically is a skill that must be learned?

What is your view of the idea that many of our opinions, beliefs, and even values, are simply adopted by us? That is, that we just inherit them without stopping to think about them for ourselves?

The Other Person's "Moccasins"

I remember seeing somewhere the Canadian native saying "Never judge another man until you have walked two moons in his moccasins." In more modern parlance, we might say that it is important to look at every situation from the respective points of view of the various stakeholders. Forgive my using the term "stakeholders" because I think it is one of those overused, clichéd expressions. It does convey, however, the importance of attempting to see the other person's point of view when we are trying to make sense of a complex situation.

Of course, this is no easy task, even when time permits. When speed of response is of the essence, it is an even more difficult task. The wise officer, however, is the one who at least makes the attempt to develop a sense of how all parties involved in an issue, and its subsequent resolution, come to see the issue and the "facts" surrounding it in different ways. Here are some questions we may be wise to ask if we are aspiring to arrive at a judgment bearing a reasonable resemblance to reality: What is likely to be influencing the accused person in terms of his explanation of the incident? What about the witnesses? Why is the shift sergeant interpreting the "facts" in the way she is? What about the district attorney's perspective? Why are there some discrepancies in the witness accounts of what they thought they saw happen? What is their relationship, if any, to the accused, and is this likely coloring their perceptions of what happened? What would I be thinking and feeling if the roles were reversed, and I were in their shoes? What does my instinct tell me about this incident, and how does this relate to the facts as they unfold? Have I allowed myself to develop tunnel vision here? Am I so convinced of the "facts" as I see them that I am simply unprepared to take a step back to reconsider my reading of the events?

Let's face it, we, as citizens, expect nothing less than absolute objectivity and open-mindedness from all parts of the law enforcement community. But in some key parts of the system—crime laboratories and forensics units, for example—those words have very special meaning. Consider the various pressures that can come to bear on a chemist or technician, because their opinions and test results will often be an essential element in making or breaking a criminal case. Needless to say, both prosecution and defense rely heavily on the honesty, integrity, and, yes, objectivity, of those lab personnel.

Several years ago, a Federal Bureau of Investigation Agent by the name of Frederick Whitehurst stepped into the public spotlight with a startling accusation. He proclaimed that the work of the FBI Crime Lab, an internationally recognized forensics section, should be looked at with suspicion since, in his view, laboratory personnel tended to fabricate test results in favor of the prosecution. Since Mr. Whitehurst was, himself, assigned to the lab, his comments had special power, meaning, and credibility.

The reaction from the criminal justice community in general (and the FBI in particular) was as expected. Mr. Whitehurst was roundly excoriated, his motives were questioned, his expertise was challenged, and his mental state was inquired into. The fact that Whitehurst's remarks tended to cast doubts on the forensics work done in the investigation of the terrorist bombing at the Oklahoma City federal building did not make things any easier, and he quickly became a pariah.

Ultimately, a number of Whitehurst's assertions were found to have merit, and new controls were put into place to ensure the objectivity of all test results. After being allowed to retire from the FBI with a substantial cash settlement to compensate for the way he was treated, Whitehurst was asked why he chose to take a position that he had to know would be provocative and unpopular with his peers. In his reply, Whitehurst spoke in terms of objectivity, open-mindedness, and walking in "another person's moccasins."

If we do become so convinced of our individual perception of what happened in a given situation, first, we are clearly not being mindful of the perceptual perils of claiming to know for sure what happened, and, second, we are covertly if not overtly claiming that we are the only individual capable of making the correct judgment!

Now I would not wish to suggest here that it is permissible or defensible to become paralyzed by the fear of making a mistake. Some people may see such careful and informed decision making as wishy-washy, or as fence-sitting. Let me disavow you of that notion. Decisions have to be made, and often they have to be made in a hurry. But jumping to conclusions based on one officer's personal perspective is generally not an approach that is going to lend itself to informed and wise decision making.

A standard check might be to ask yourself something like "How would I feel in this situation if I were this person? Am I looking at the entire situation, and am I responding to it in as objective a way as possible?"

Attempting to put ourselves in the other person's moccasins might be one step in the right direction. By so doing there may be an enhanced chance that justice will be served.

Reflections

Can you think of any case where justice was not served because the investigating officers jumped to a premature conclusion about an accused person's guilt or innocence?

If you can think of such an incident, might the process described here of attempting to see the situation from many perspectives have helped to bring about a more appropriate and less embarrassing conclusion?

Moral Stages

One important figure in the area of moral development and moral education is Lawrence Kohlberg. I would encourage you to look up some of Kohlberg's work, and to engage your facilitator in a discussion about his ideas.

It is my hope that Kohlberg, although now deceased, will forgive me for reducing what can sometimes be fairly complex theories of moral development down to what can only be described as the most simplistic of levels. His work advocates that humans operate at any one of six stages of moral development. The stages range from 1, which is the least morally developed and principled, to 6, which is the most developed and highly principled stage.

If we are at stage 3, for example, he claims that the motivation for our behavior comes from a need to conform, where relationships are critically important to us and govern our moral decision making (ring a bell?). Stage 4 is the stage at which we make our choices based on what Kohlberg calls "conscience maintenance." If we don't feel guilty, we must be doing all right. Stages 5 and 6, on the other hand, are stages at what he calls the "post-conventional" level, where individuals are able to make reasoned moral judgments based on sound ideals and ethical principles, and where those individuals are more likely to have the moral courage to follow through on their moral choices.

Now what is the point of all this? (And I am sorry if this all sounds convoluted!) The point is that if we think of stages 3 and 4 as basically "going with the flow," and stages 5 and 6 as principled and courageous, we may ask ourselves an important question just prior to making a moral choice: If I make the moral choice I am about to make, at what stage might I be operating? Is my choice based more on what others think of me and what I think of them, or is my choice one that I can truly call my own and that is reasoned, principled, and strong?

It's a tough question, but it can be a useful guide or check on our behavior.

Reflections

At what stage do you think you generally operate in the ethical choices you make?

What do you understand from the term *moral courage?*

Putting the Tools to the Test

Let's have a go at working through a moral dilemma using the tools. We will do so briefly, but the process may take a little longer in class because of the diversity of opinion that no doubt will be brought to the issue.

As this chapter was being written, I had lunch with a chief of police, and during our chat he related a dilemma he had been thinking about. He said, "Geez, I've got this problem. We have some neighbors several doors up from us, and they are absolutely delightful people. We have known them for several years." He explained that the two families had socialized together. "Their son is a graduate from a college justice program, and desperately wants to get on with my force. But you know, I kind of have some reservations about him. There's something that I can't put my finger on, I feel kind of uneasy about him. Obviously his parents would like me to help because the kid has got his heart set on policing, but I don't know."

So, briefly, let's try to apply the tools to the problem. The dilemma is, should the chief help his neighbor's son to get hired by his police department?

The first tool, setting limits, appears to have little or no direct application to this problem.

The second tool is ethical principles, the first of which is the principle of benefit maximization. Now if we apply this principle to the problem, we quickly see that the greatest benefit, if the lad were hired as a result of the chief's influence, would accrue to the son and his parents. The fact that there may be a much better candidate available would mean that the interests of the community would probably not be well served. Overall, therefore, it is likely that less public good would result if the son were hired. The second principle, equal respect, refers to giving all individuals an equal amount of consideration. That would clearly not be the case if the chief gave this lad an edge by exerting his influence. So far, then, the tools would indicate that the chief should not use his influence to help the neighbor's son.

The next tools we discussed were the bell, the book, and the candle. In this instance, I am unsure about what the chief, if he were to use his influence to help this lad, would say about the bell. Only he would know. The book, however, may be helpful. Police force policies would most certainly address issues of conflict of interest, the importance of fairness, and the inappropriateness of senior officers using influence for personal reasons. (This may mean, too, that warning bells would be going off.) The candle, referring to whether a decision will stand up to public scrutiny, would indicate that the chief should not use his influence in this case. Clearly, most members of the public would wish that police hiring decisions would be based on fairness and equal opportunity. So far, so good: The chief would be unwise to use his influence.

The next tool, the imaginary video camera, would probably also indicate that it would be unwise of the chief to help his neighbors, because if his use of influence in this way were recorded or documented, he would probably have to answer some hard questions. The message continues to be that he should not help the lad any more than he is prepared to help any other candidate.

Critical thinking is the next tool. If the chief is prepared to put his decision making to the test he has already submitted himself to, he has begun to engage in critical thinking. He would now be beginning to see the issue much more clearly, and the decision-making process would be removed from the gut level, to a level where neighborly loyalties are probably not going to dictate the decision.

When we consider the chief placing himself in the other person's moccasins, an important question to ask might be "If I were a candidate for this police force and I thought I was placed at a disadvantage because of a senior officer's 'pull,' how would I feel?" The answer would appear to be self-evident. Conclusion: The chief should not use influence to help the neighbor's son.

Finally, can consideration of Kohlberg's moral stages help us here? Well, if the chief wants to be liked, and wants to reinforce to an even greater extent his neighborliness with the lad's parents, he can help the lad. But would this be a good reason? Perhaps a better way might be to think the whole thing through independently, and then make the decision that is the right one, irrespective of any potential fallout with the neighbors. Loyalties, therefore, would not be permitted to "color" his thinking about the situation, nor to dictate his decision.

It would appear that the chief can make his decision with confidence and move on. The agonizing is over. To use his influence in this case would probably be unfair, and would probably not withstand public scrutiny. And perhaps most important, the general good of the community would be better served by hiring the best candidate available.

My hope is that this little exercise will demonstrate how our moral decision making can be helped by submitting the dilemmas we encounter to a similar process.

Summary

Perhaps the acid test underlying all of these "checks" is this: How do I feel about myself? Do I feel strong? Or, do I feel awkward and ill at ease about my action? Do I feel shaky inside? Am I depressed about the choice I made? Do I find myself constantly looking for ways to justify my choice?

I think it would probably be safe to say that few of us, and I include myself here, have not experienced the soul-searching indicated by troubling questions like these.

The biggest worry of all, however, would be the person who has never felt this struggle, for none of us is perfect. The individual who has experienced the struggles can, at least, take some comfort from the fact that her moral sense is still active.

We are human, and part of being human means that we will make mistakes.

However, the mistakes may be fewer, and the really big, career-destroying and peace-of-mind-destroying mistakes may be fewer, if not eliminated, if we adopt some of these techniques to help us work through the moral dilemmas confronting us in our work.

These tools may help us think more clearly, and they may serve to reinforce our moral sense, but in the final analysis the entire issue boils down to one of moral courage. I know what I should do, because I have thought about the dilemma carefully and honestly, but do I have what it takes to follow through with what I think I should do?

Moral fortitude does not come easily or readily.

Let me leave you with one final thought in this chapter. Might it be the case that the courage to act may come more readily for having systematically, honestly and thoughtfully weighed the consequences of our decisions?

Where Do You Stand?

There is always more than one choice to be made. I do not think there are right and wrong choices, but rather better and not very good choices.

—Journal entry, justice student in ethics course

I am a strong believer that when people choose to work in the justice field that the most important factor is honesty. You must be honest to yourself as well as others.

—Journal entry, justice student in ethics course

I believe that if a person thinks in a certain way then he/she should be able to give reasons to support why he/she thinks that way.

—Journal entry, justice student in ethics course

I have always found it curious that as a society we discourage people from changing their minds. How did it become a sign of weakness and not strength? For me one of the most admirable qualities in another person is the ability to say "I changed my mind, I never thought of that before." In my view, that indicates strength and intelligence and open-mindedness.

—Journal entry, justice student in ethics course

Introduction

As you can see, this chapter has no Chapter Objectives section. The purpose of this chapter is simply to give you some material for your own thinking, and for discussion with your classmates.

You may remember my stating in Chapter 1 that the dilemmas included in this chapter are based on actual experiences. They are not hypothetical. They have been directly or indirectly recounted to me by officers in the field, so I hope they will ring a few bells for you.

The police and corrections scenarios are not separated, and this has been done deliberately. You are encouraged, whatever criminal justice profession you are studying for, to have a go at working through several, if not all of the issues. In this way you will develop some insights into the kinds of awkward problems that can confront both police and corrections officers.

When you are thinking about the dilemmas and how you or your colleagues would probably respond to them, attempt to place yourself in the actual situation. It is one thing, in the comfort of one's home or the relative comfort of a classroom, to imagine how we would respond. But I want you to try and transport yourself into the situation and imagine the dilemma unfolding. Try to grasp all of the nuances of the situation: the subtle and not-so-subtle pressures, the friendships, the comradeship, the expectations you have of yourself. Think about who you are as a person. Think about the faith your parents and family probably have in you. Look at each situation in its entirety and attempt to weigh all the factors before you begin working out what you think you would do.

Let me offer a word of caution. If you read a dilemma and you think you know immediately what you would do, there is a good chance that you have not grasped the complexity of the issue. They are dilemmas because they are likely to induce in you a similar response to the one experienced by William Tafoya's subject: "I guess I was damned if I did, and I was damned if I didn't."

Please also be aware that thinking we know what we *should* do in a given situation, and thinking we know what we *would* do in that situation may be two different things. If there is a discrepancy like this between your thinking and your anticipated behavior, what might account for the discrepancy? Why is it, do you think, that at times even when we know fairly certainly what we should do, we can find ourselves lacking the moral courage to follow through?

So here are the dilemmas. Some relate to police officers, others to correctional officers, and others to both professions. Several scenarios ask you to consider dilemmas facing senior officers, so that you can begin to realize how difficult it can be at times for administrators when they, like you, have to wrestle with what should be done.

You are not being asked, however, to wrestle with these dilemmas without help. I would recommend that you go back to Chapter 2 and review the discussion there before you attempt to work your way through the dilemmas. Look at what we had to say about the nature of a moral dilemma. Remind yourself about what we said about gullibility and mindlessness; about what Dr. Martin Luther King had to say about good laws and bad laws. Refresh your memory regarding what we meant by libertarianism, determinism, utilitarianism, and the principle of equal respect. Think again about what Plato, Socrates, Aristotle, Hobbes, and Sartre had to say about justice. If you learn to incorporate these concepts and ideas into your thinking, you may find that your thinking will become clearer and more comprehensive. May I also suggest that you practice using the tools discussed in Chapter 8. Have fun with the dilemmas. But at the same time remember that one of these days you may find yourself in a similar situation for real. I hope the experience of thinking both privately and publicly about them, and discussing these dilemmas with classmates, will help you when the time comes.

You should not feel "led" in any direction by the questions. They are not designed to move you toward thinking in a prescribed way about an issue. They are included simply as prompts for thought and discussion. What matters most of all is what you think and feel about these issues, whether or not you can defend your position on them, and what reasons you employ to support your response.

Try to remember what I said in an earlier chapter about moral courage beginning when we are willing and able to take a hard, honest look at dilemmas like these; and when we are strong enough, at least for a few minutes, to put aside our prejudices, biases, and preconceived notions, and honestly face up to the hard questions.

Enjoy the experience. You may learn something important about yourself, your colleagues, your chosen profession—even life in general!

Dilemma No. 1

Is Honesty a Personal Quality—Like Shyness?

When I was department administrator of the justice department of the college where I work, two police science students were brought to my office by a faculty member. They had been caught cheating on an examination.

I asked them both if they had, in fact, cheated. A flood of excuses followed, accompanied by interesting you-go-first glances from one to the other. It was only after the cheat sheet was produced and clearly matched one of their answers that they grudgingly admitted to cheating.

Then something interesting happened. One of the students became quite emotional, admitted his guilt, said he was ashamed of himself, and said he had learned a painful lesson from the experience.

The other student, on the other hand, continued to be defiant, insisting that many students cheated on examinations. I asked him if he thought that honesty was an important personal quality for a police officer, and he replied, "Well, it would be different if I had a uniform on." There is a curious logic to this response.

Both students were asked about their career aspirations. The defiant student said he wished to join a state police agency, and the other student said he wanted to join a federal organization.

Questions

1. Is being honest a personal quality or characteristic—like shyness, for example—or is it something we simply decide to be, depending on the circumstances in which we find ourselves?
2. Is there any moral difference between cheating on an examination and stealing from an unlocked electronics store while on night patrol as a police officer?
3. If you were me, and you heard six months later that the defiant student was in the final stages of the recruitment process for a state police agency, would you, or should you, pick up the phone and give the recruitment section the information you have?
4. If your answer to Question 3 is no, why not? If yes, why?
5. Is a college professor or administrator's primary responsibility to his graduate student or to the general welfare of the community?
6. If a person appears subsequently to be genuinely remorseful for a wrongdoing, should this be a mitigating factor in the way the situation is dealt with?
7. Imagine a situation at a police or corrections training center where in, say, the tenth week of a 12-week orientation course a student is found cheating on an examination. Would this circumstance warrant, in your view, dismissing that trainee officer from the force?
8. Which of the tools, if any, may help you with this dilemma?

Dilemma No. 2

Thank God for Sick Days

(*Note:* This is a dilemma that can be applied across the board in the criminal justice field. Please be aware that this scenario has the potential to put you on the defensive. Please try to look at the situation as calmly, logically, and reasonably as you can.)

South Side Detention Center is a maximum security institution housing inmates on remand or serving sentences of 90 days or less. It is a government facility, so the staff are public servants.

Recently there have been several reports in the newspapers about the troubling levels of sick leave used by public servants. The average number of sick days (or "sickies" as they say in Australia) taken during the last year by government workers has been reported and confirmed to be an average of 11 for each employee. In correctional institutions and police forces the average number is 15 per employee. The cost to the taxpayer is estimated to be tens of millions of dollars each year because of the overtime or substitute staff needed to provide shift coverage.

The attitude of many of the staff at South Side Detention Center is that the sick days are one of their benefits. They feel that sick days are provided for in the contract and are there to be used, whether they are sick or not. "If you don't use them, you lose them," appears to be the prevailing attitude.

You are an officer at the detention center, and one day you are sitting in the lunchroom and you hear the following conversation.

Officer 1: I owe this place nothing. I make no bones about it, if I want to use one of my sick days for a day off I use it, that's all there is to it!

Officer 2: I know I haven't been here long, I don't know too much about what goes on here. My understanding was that those sick days are there for us in case we really are sick.

Officer 1: That's the way most of us felt at one time, but not anymore. If I want a day out with my wife, or a day out fishing, I'm out of here on a sick day.

Officer 2: I'm not sure I would be able to relax. I think I would feel guilty all day long, so I may as well go to work. Besides, what if you get caught?

Officer 1: So what? The boss does it too. We all do it. It's a right. It's in the contract.

Officer 2: I don't mean to sound self-righteous or anything, but isn't it a form of stealing? If I phone in sick when I am not sick, if I go fishing on a sick day, I am being paid to go fishing, aren't I? I am not sure about that.

Officer 1: It can't be wrong because everybody does it.

Questions

1. Is this problem primarily an economic one or an ethical one?
2. Is it morally defensible to go fishing on a sick day when I am not sick, but when that activity will probably enhance my mental health?
3. Does a manager or senior officer have any more of a responsibility to act morally than a frontline officer has?
4. Is Officer 2 in the dialogue responsible only for her conduct? If she becomes aware of an abuse of the sick leave policy by a colleague, is she morally "on the hook" to do something with that knowledge?
5. Some people say that silence implies consent. What do you think?
6. Let's assume for the sake of argument that a sick day is worth $200. If I take a sick day when I am not sick, isn't this something like taking $200 out of petty cash? If not, why not?
7. Which of the tools, if any, may help you with this dilemma?
8. To what extent, if any, do you think your response to this kind of dilemma would be governed by officer subcultural constraints?

Dilemma No. 3

The Timing Stinks!

The Mount Henry Police Force (the name is fictitious) is a small, municipal force in a semirural community. The force has a chief, deputy chief, five sergeants, and 37 sworn officers. Generally, the force is considered to be well run, and enjoys the respect of the community it serves. The chief is considered to be a politically active individual, and is on good terms—is friends, even—with almost all of the local officials.

It is a damp, cold November evening and Officer James has parked his car in a rest area a few miles from the station to get caught up with his paperwork. (James is a probationary officer, with just three months of his 12-month probation remaining.) The roads are quiet and he is enjoying a few moments of peace to collect his thoughts, and to reflect on what has been a long shift.

At that moment the police car in which he sits is illuminated by the headlights of an oncoming car. The headlights are on full and dazzle James for a moment. The car passes him slowly, and when he turns around to take a better look at it, he notices it is being driven erratically. "Thank God," he thinks to himself, "there are no other cars around."

He starts the cruiser, turns on his lights, and swings around in pursuit of the car. As he approaches the car, he notices it is increasing in speed. It veers alarmingly from side to side, and the young officer becomes increasingly anxious at the thought of what might happen if another car happens to approach from the opposite direction.

He accelerates, catches up with the now speeding, veering car, and turns on his emergency lights. Gradually the citizen's car slows down and stops on the shoulder of the road.

Shaken, he gets out of his cruiser, approaches the vehicle—and shines his flashlight into the face of the chief of police. The chief explains that he had been to a difficult budget meeting, and afterwards knocked back a few drinks. He had assured his friends, who expressed concern for his safety before he left, that he was fit to drive, especially since it was not far to his home, and along a quiet road.

"Officer James, you're a good man. My house is just up the road; I'll be fine. Good work. I'll remember you for understanding the situation. Those jerks at City Hall don't understand what these budget cuts are going to mean to us. Anyway, whatever happens, even though you are the youngest officer on the force, you have my word that you will not be affected . . . if you see what I mean."

Questions

1. What should Officer James do?
2. What would you do if you were in a similar situation?
3. Officer James has to think about his wife, his baby, a mortgage, and a career he loves. Should his decision be influenced by these responsibilities and concerns? Where should one's priorities lie in situations like this one?

4. Keep in mind your responses to Questions 1, 2, and 3. What if you were a police officer pulled over for drunk driving, and the investigating officer, a colleague and friend of yours, chastised you for being a stupid #$*!, but said he was going to give you a break. Should you—would you—accept his kind offer?
5. Should the law be applied in one way for the police "family" and in another way for ordinary citizens?
6. Which of the tools, if any, may help you with this dilemma?
7. To what extent, if any, do you think your response to this kind of dilemma would be governed by officer subcultural constraints?

Dilemma No. 4

I Guess He Should Have Stayed Home

Ted Egan has worked at the county jail for 13 years. He is a dedicated, competent employee. He is well respected by his colleagues and by the administrators of the institution.

Several times during the last 10 years Egan has been approached by senior officials to encourage him to apply for promotions as they became available in the institution. Each time he declined, explaining he was happy doing what he was doing because it meant he could spend more time with the inmates.

But lately things have not been going very well for Egan. His wife of 21 years has just announced she is filing for divorce. She has met another man. In addition, Egan's 13-year-old daughter is acting rebelliously. Egan, particularly because of the business he is in, is keenly aware of a serious breakdown in communication between himself and his daughter. The marital problems are certainly not making matters any easier.

Now Egan is at work. It has been a difficult shift on this particular day. He reported for work at 7.00 a.m., after an emotionally draining evening and night dealing with problems at home. Clearly, he arrived at work tired and worried. As is often the case at stressful times like this, the inmates seem to sense he isn't his normal self, and start to act out.

One particular inmate is getting on Egan's nerves. He is 18 years old, small and "yappy." He never stops. For several hours he has been carrying on with his high-pitched, pointless, and increasingly maddening "yap."

Egan somehow manages to remain in control of himself; that is, until this particular individual comes into the unit office to complain about some of the other inmates. They had told him in no uncertain terms to "shut up or else." At this moment Egan loses control and slaps the inmate hard across the face. The blow is hard enough to leave a red welt and a small cut. The noise of the blow is loud enough to be heard by the other inmates—who spontaneously break into a round of applause! In 13 years this is the first time that Egan has lost his temper in this way.

Another officer is called to cover Egan's post, and he goes to report the matter to the superintendent. She invites Egan in, and listens intently to someone she normally admires and respects as he recounts the incident.

"I did it," he said. "I hit him hard. He just got to me. Things have not been good at home lately—but that's no excuse. I know you have to fire me. Or would you let me resign? I'm truly sorry it had to happen this way."

Questions

1. If you were the superintendent what should you do? Should you fire Egan? Should you allow him to resign? Should you perhaps take a course of action that is quite different from these two?
2. Should an officer, in a position of public trust, be fired for making one uncharacteristic mistake when there appear to be extenuating circumstances contributing to that mistake?
3. Should the superintendent, in making her decision, take into account the probable effect of Egan's dismissal on the morale of his colleagues? Or is this simply not an issue?
4. Assume that the superintendent made a decision not to fire Egan; let's say in this instance he would be warned and reprimanded. Might the superintendent be in danger of signaling that it is okay to engage in unprofessional conduct, as long as it occurs only infrequently?
5. Should Egan's personal circumstances, and his subsequent attitude of contrition, be a factor in the superintendent's decision?
6. Which of the tools, if any, may help you with this dilemma?
7. To what extent, if any, do you think your response to this kind of dilemma would be governed by officer subcultural constraints?

Dilemma No. 5

Sleeping Arrangements

You are a relatively new correctional officer at the Orangeburg Correctional Facility. You have worked at the institution for nine months, so there are three months remaining of your one-year probationary period. At that time, with satisfactory reports from supervisors and colleagues, you will earn permanent status.

Orangeburg is a maximum security facility housing male offenders, most of them in jail for having committed violent offenses.

You are working a string of seven night shifts that run from 11.00 p.m. until 7.00 a.m. This is the sixth of the seven shifts and you are feeling exhausted. You are not finding it easy to sleep in the daytime, partly because you are not yet used to it, but mainly because your one-year-old daughter is now quite active and noisy.

On this night you are assigned to work with an officer who has worked at Orangeburg for the last 17 years. At 2.00 a.m., he approaches you and tells you that he is going to take the next two hours for a sleep, and that you should go on your rounds without him. At 4.00 a.m., he says, you should wake him up, and then it will be your turn for two hours of rest. You are aware that this is strictly against the standing orders of the institution, and that being caught asleep by a senior officer is grounds for dismissal.

You explain to your partner that you feel uncomfortable with the "arrangement," and you insist, for your own safety and the general safety of the area you are patrolling, that you do the rounds together. Your partner becomes highly agitated, asks you who you think you are, waves his finger in your face, and tells you that you will do as you are told. After calming down a little, he explains to you that he has an important event to attend the following morning, and that he has been banking on getting some sleep during the shift. He reminds you that you are on probation, and that if you think you can do anything about the conversation you have just had, you may as well forget it, because he would be believed and supported over you.

Questions

1. How do you think you *would* handle a situation like this one?
2. What do you think you *should* do?

3. Is it ever ethical to tell one's supervisor about the poor performance or objectionable behavior of a fellow officer?
4. Let's assume that during the following day you make further surreptitious inquiries about your predicament with someone you trust. You discover that the lieutenants are also in the habit of taking their turns to sleep. Now what?
5. To whom are you primarily responsible:
   Your colleague?
   The inmates?
   Your employer?
   Yourself?
   All, some, or none of the above?
6. Some people are of the view that if one officer knows of another officer's serious indiscretion, but chooses to stay silent, both officers share responsibility for the indiscretion. What do you think?
7. Which of the tools, if any, may help you with this dilemma?
8. To what extent, if any, do you think your response to this dilemma would be governed by officer subcultural constraints?

Dilemma No. 6

The Air Drop

A sense of excitement is running through the members of a multiagency drug task force this evening; in a couple of hours it looks like a lot of hard work is going to pay off. After seven months, lots of leg work, many hours away from the home and family, and extensive interactions with informants, arrangements for the air drop of drugs finally appear to be in place. Several members of your team are already positioned around the area where a small plane, flying low and without lights, is scheduled to drop the contraband. You and your remaining colleagues will be situated nearby, ready to move in to make the capture after the packages are picked up. Things are getting tense now; an informant in the country where the drugs are coming from has just let you know the plane has taken off and is on the way.

At the appointed time, a small plane flies over and several packages fall to the ground. The observation team sees a man move out from cover and begin picking up the packages, and you and your team move in for the arrest taking a very surprised drug recipient into custody. There is, however, a palpable sense of disappointment in your group, for there had been an expectation that some individuals from higher echelons in the drug trade would be present. But that was not the case, and the lone arrestee is being questioned about the identities of others involved in this transaction. So far, though, he is not yielding any information whatsoever.

As the questioning continues, you notice that a lieutenant from a different agency is beginning to lose patience. "Are you absolutely sure you don't have any names for us?" he asks. "Nope," says the defendant with a shake of his head. "One last time," the Lieutenant says. "Do you have any names for us?" With a sneer, the defendant repeats "Nope."

The lieutenant grabs the handcuffed man, drags him toward a police vehicle parked nearby, opens the trunk of the car, takes out a length of rope, and ties one end of the rope to the handcuffs. He then ties the other end of the rope to the rear bumper of the vehicle, and says to the man, "Last chance. Any names for us?" Though considerably more concerned by this point, the man still replies in the negative. The lieutenant then turns to another officer behind the wheel of the car and tells him to "take off." The car pulls away, knocking the arrestee to the ground and pulling him behind. After being dragged about 50 feet, the defendant screams "OK, OK. I'll give you the names." He is released from the rope and handcuffs, and gives your team the names of four high-ranking members of the local drug organization, and the location where substantial stores of illegal drugs are being kept.

Questions

1. What should you do here? Does it matter that the lieutenant outranks you? Does it matter that he is from an agency different from yours?
2. No other members of the multiagency drug task force appear to be bothered by these events. Does this affect how you might choose to respond?
3. Illegal drugs are a scourge of our society. Is there a "greater good" involved here? Is it OK to apply physical force in order to obtain information that will protect the community?
4. Would your feelings about this change if you were a member of a multination drug task force and you were in a foreign country when this occurred? What if the lieutenant were a member of that country's national police force?

Dilemma No. 7

The Good Friend

As a patrol officer working the day shift in the summer time, you can't wait to get off duty, take off your sweaty uniform, and go home to your family. When you arrive home this evening, your wife greets you at the door with a cold glass of lemonade, and tells you that a friend of yours called earlier this afternoon and would like you to call him back. This friend, who works in the computer industry, has served on a number of church committees with you over the years; in fact, you and he currently serve on the board of trustees at your church. He and his wife have two children about the ages of your own, and your two families have socialized together on many occasions.

When you call him, he says that he has something important to discuss, and asks you to meet him for coffee tomorrow. You are off duty the next day, and you agree to meet him at a coffee shop in a nearby town, different from the one where you live and are employed as a police officer.

When you sit down at the table with your friend he looks at you, pauses for a moment, and says "I called you because you are a friend, and I need someone I can trust to help me work through a problem." He then goes on to tell you: "Last week I was arrested for exposing myself. I am absolutely humiliated; in fact, I haven't even been able to tell my wife yet. And the fact is . . . I did do what I was charged with."

Questions

1. What should you do? Does your friend have an expectation that you will keep this information to yourself?
2. The arrest took place in a community different from the one in which you work. Do you have a legal obligation to report your friend's confession to the police in that community? Do you have a moral obligation to do so?
3. As a police officer, what are your responsibilities in this matter? As a friend, what are your responsibilities in this matter?

Dilemma No. 8

Lies, Internal Affairs, and Videotape

As soon as the woman walked into the Internal Affairs office, two things were very clear: (1) She was angry and (2) she wanted to make a complaint. According to her, she had been treated badly last Saturday night after she was arrested for driving under the influence, and she wanted something done about it. She said she had been manhandled and abused while being processed into the city jail, naming both the arresting and booking officers as the targets of her ire. Specifically, she alleged that when she was in the booking and processing room, both of those officers yelled at her, used profanity, called her obscene names, and pushed her against the wall and onto the floor several times.

The Internal Affairs investigator knew this would be an easy complaint to resolve, since all booking processes are automatically videotaped for just these purposes. After taking the complainant's statement, she went to the booking area, retrieved the tape containing the events of the last Saturday night, and fast forwarded to the booking process at issue. The tape showed that contrary to allegations, the two officers had treated her with the utmost courtesy and professionalism. In short, the complaint was unfounded.

In locating that section of tape, though, something else caught the investigator's eye, and she rewound to that point. What she saw was almost unbelievable, for it showed two officers and a sergeant beating a man with nightsticks about four weeks earlier in that same booking area. A full viewing showed that the citizen (another DUI) had been brought in after being involved in a serious traffic accident, and he was highly intoxicated, uncooperative, and combative as well. It appeared that the officers eventually had lost patience with the drunk, at which point they began to beat him and continued to do so even though he had stopped resisting after the first baton strike. When she checked on this event, the investigator learned that there had been no complaint made by the assaulted citizen. He had been released on bail two days after the video had been made, and was awaiting his first court appearance in about one week on the DUI charge. No other charges had been placed against him.

Questions

1. What should the Internal Affairs investigator do about this? After all, there has been no complaint made.
2. This apparent assault took place about four weeks ago, and the videotapes of the booking area are erased after 30 days. Soon there will be no trace of this ever having taken place. Should the investigator simply wait until the evidence disappears?
3. The citizen may have assumed any injuries he sustained in the beating occurred as a result of the traffic accident he had been involved in. Do you think the videotape might refresh his memory about what actually happened?
4. If this information is released to the citizen and made public, it is likely the city will be forced to pay a settlement in some subsequent lawsuit. What do you think the chief of police would want done with this?

Dilemma No. 9

The Racist Recruit

Thank goodness it is almost over. Sometimes it seemed like the Basic Recruit Academy would never end, but now there are only three weeks until graduation. There are 35 recruits in the class, and gradually all have gotten to know a bit about one another. Like every group, some are more quiet than others, but everyone seems to have gotten along pretty well.

In a five-person study group this evening, one of the quieter recruits has begun to open up more than usual, and with the elevated sense of comfort, a new facet of his personality has come to the surface. You are studying together in the room of one of the other recruits, and during the course of the one-hour session, this fellow has used a number of racist and sexist terms to describe other recruits and members of the public with whom he will soon be coming into contact. Because he and all the other members of the study group are white males, perhaps he thought his views would not be challenged. He may have been correct, for with the exception of your comment that he should "knock it off," nobody else spoke up.

When you return to your room, you can't shake the notion that this recruit's behavior is a problem. You walk down the hall to the room of one of the other recruits from the study session and share your concerns with him. He tells you to "forget about it," especially since the academy is almost finished, and chances are you will never have to work with this guy out in the "real world."

Questions

1. What should you do about the racist and sexist remarks made by your fellow recruit?
2. Does it matter that nobody else in the group appears to have seen this as a problem?
3. If you report this, will there be any repercussions for the other recruits who were in that session but who did not report it? Should there be?
4. Would it make a difference if you knew the recruit who made the questionable remarks was married with two small children, and if this gets reported he might very will be fired?

Dilemma No. 10

The Drunken Refusal

When he saw the car weaving across the center line of the highway, Mike had a strong suspicion that it was being operated by a drunken driver. Calling for a backup, he turned his emergency lights on, and gradually the civilian car eased onto the shoulder of the road. When he approached the driver's side door, the odor of an alcoholic beverage was very strong, so he asked the motorist to step out of the car and join him and his backup officer on the side of the road. Because the man appeared intoxicated, Mike decided to use a portable breath test device to confirm his suspicions. He inserted a sterile plastic mouthpiece in the device, gave the motorist instructions on how it worked, and instructed the man to blow into the mouthpiece. As sometimes happens, the fellow attempted to thwart the test by appearing (but failing) to blow into the mouthpiece. After giving the man several opportunities to do it correctly, Mike yanked the mouthpiece out of the device, threw it on the ground, and arrested the man for DUI and refusal to submit to a breath test.

Several weeks later an administrative hearing was held to determine whether the motorist's license should be revoked because of his refusal to submit. At that hearing, the driver's attorney told the hearing officer that his client had, in fact, tried to submit to the test, but the plastic mouthpiece was flawed (blocked) in some way, so he was unable to blow through it. The hearing officer turned to Mike, who was under oath, and asked him whether the mouthpiece had been blocked. Mike testified that the mouthpiece had not been blocked, and as evidence of that fact, he took a mouthpiece out of his pocket and handed it to the hearing officer, stating that it was the mouthpiece he had used for the test that night.

The backup officer happens to be sitting in the room during this administrative hearing, and hears Mike testify to the fact that the mouthpiece he presented to the hearing officer is the same one that was used that night. He knows differently, because he saw Mike throw that one on the ground after he became impatient with the motorist.

Questions

1. What should the backup officer do about this?
2. Does it make sense to risk one's career over the particulars of a DUI arrest?
3. If Mike is willing to perjure himself on a matter like this, can his word be relied on in any other matter in which trust is involved?

Dilemma No. 11

To Care or Not to Care . . .

This is the sort of dilemma sometimes wrestled with by police managers. It is included here to give you a sense of what it can be like to wrestle with a manager's dilemma, to get you to put on the other person's moccasins, as it were.

Late in 1997 I attended a police ethics conference in Dallas, Texas, hosted by the ethics center at the Southwestern Law Enforcement Institute. There were many thought-provoking sessions but one, more than any other, stands out in my memory.

The presentation was being made by a police officer who had himself been deep undercover for the previous five years. His presentation, delivered without much polish or fuss, simply asked the question: "Do police managers have a duty to take care of, and be supportive to, officers who, because of the peculiar and high-risk nature of their work, suffer personal damage as a result?" The presenter used his own experience as an example of how deep undercover work can "screw up an officer's thinking and outlook on life." "At one point in your life," he said, "you have access to unlimited sums of money provided you by the force. There is a very little accounting for these large sums of money. You are leading a life where gratuitous sex, booze, and drugs are yours for the asking. And you are expected to play your role very convincingly. If you don't, you could die. And then, all of a sudden, you are pulled out, and you are expected to go back to regular duty as though none of the previous five years happened."

The officer made it quite clear to those present that he had paid an enormous personal price for his undercover work. His personal relationships had suffered, his values had been severely challenged, he suffered all kinds of role confusion, and he was severely stressed by his experience.

Permit me to add a footnote to this. During my years of traveling around conducting seminars, particularly ones on suicide prevention and hostage survival skills, I have experienced officers who break down and cry about incidents they had experienced in the course of their work. One officer broke down about a suicide he had failed to prevent 18 years earlier; another cried about storming in the middle of the night what turned out to be the wrong house, and roughing up an entirely innocent family; one cried about ignoring deeply troubling signs of distress in a colleague, thinking he was doing his colleague a favor, until his colleague hit a career-ending jackpot; one cried when he recalled the details of his first fatal accident involving a four-year-old child.

Due to the nature of police and corrections officer work, individual officers can, from time to time, pay a deep personal price for the duties they may have to perform during the course of their work.

Questions

1. Is it primarily a practical or primarily an ethical question that managers should respond compassionately to officers harmed psychologically or emotionally during the course of their duties?

2. Should middle managers, say, sergeants and lieutenants, be especially careful about showing too much compassion in case the troops become "soft?"
3. Should the admission of psychological and/or emotional stress by an officer (after a serious incident) be a factor in future promotional considerations?
4. Would submission to counseling or to critical incident debriefing by a colleague officer cause you to consider that colleague "suspect."
5. Would your response be different if you were talking about a female colleague officer?
6. One police officer in New South Wales, Australia, went to his sergeant and said, "Can you please take me off highway patrol for a bit, because I am feeling stressed out by what I'm seeing out there." The sergeant decided to "toughen the officer up" and sent him out to patrol the most dangerous section of the Pacific Highway, a two-lane highway running north from Sidney. Several weeks later, this officer was the first on the scene when two buses collided head-on. Almost 70 passengers were killed and the ensuing carnage was beyond imagining. Three weeks later the officer went into a meadow, took out his service revolver, and blew his brains out. Should or does the sergeant carry any moral responsibility for the police officer's decision to kill himself? If yes, why? If not, why not?

Dilemma No. 12

It's Rough Justice, But If It Works, What's the Problem?

You work as a corrections officer in John Street Detention Center. There are 15 young offenders in the unit. Many of the offenders have been placed in your unit because to lesser or greater degrees they manifest mental health problems.

Even though you have worked at the facility for only two months, you feel quite comfortable. The long-term staff have accepted you, include you in their conversations, and they are beginning to invite you to their social activities. Other than one or two minor difficulties you have with some of the inmates, which is quite normal, you have experienced no real problems and life is generally good. You admire the skills of the experienced staff and so far you have learned much from them in how to handle difficult situations.

It is a Tuesday at lunchtime. You are in the dining room and you notice eight of the inmates are sitting at one table. One of them is a very immature 18-year-old (perhaps with a developmental disability, but you are not sure) called Peter. His manners at the meal table are disgusting. In fact, after seeing Peter eat you are generally left with no appetite because his eating habits are so off-putting. One habit in particular upsets you and others—he pours mounds of ketchup onto virtually every meal, swirls it around on his plate, and then slurps it into his mouth. Sometimes he uses a spoon but as often as not he uses his hands. He then eats with his mouth open and often spits the food out across the table when he's talking; burps often follow. You, the other officers, and even the other inmates whose own eating habits are not that great, are sickened by his behavior.

This particular lunchtime Peter's display is especially *bad* and it's clear that everyone sitting at the table is feeling nauseated.

A fellow officer, Tom Castle, surprises you when he gets up and tugs Peter away from the table by his shirt collar. They disappear into the kitchen and Tom comes out with a deep-sided bowl. He orders Peter to scrape what's left of his meal into the bowl, takes Peter to the center of the dining room, places the bowl on the floor, and tells Peter to eat his meal like a dog. "Your manners are disgusting," said Tom. "You are making us all feel sick and if you are going to eat like a dog you may as well get down on all fours like a dog. Get down on your hands and knees—you can stay there until the bowl is licked clean."

Tom later explained to you and the other staff member present that while he was acting out of frustration, he did what he did mainly as a shock tactic to get the message through to Peter. In this way Peter would be less likely to alienate those around him, which would make his stay at the detention center less problematic for him in terms of his relationships with the other inmates.

Questions

1. What, if anything, would you do in this situation? What, if anything, should you do?
2. Do you think Tom acted professionally? Did he act ethically? Is it ever defensible to use somewhat dubious behavior modification techniques to eradicate socially unacceptable behaviors?
3. Assume you were offended by Tom's response, that his response was over the top and indefensible. Would you have intervened on Peter's behalf?

4. What should you do, if anything, if you were the superintendent of this facility and this incident was reported to you?
5. To which of the tools would you appeal to justify your arguments?

Dilemma No. 13

How Much Is One (Crazy) Human Life Worth?

This is a dilemma that has to do with one particular stand-off that took place in a rural, farming community called Roby, Illinois. Providing substance for the write-up of this dilemma is an excellent article written by Inspector Richard E. Dunn of the Illinois State Police called *Ethical Police Behavior at Roby Ridge.* The article was published in *Command,* an official publication of the Illinois Association of Chiefs of Police (Spring 1998, Volume 8, Number 1). Also providing background is a talk on this subject given by Terrance W. Gainer, now executive assistant chief of police, Metropolitan Police Department, Washington, DC, (then chief of the Illinois State Police) at a police ethics conference I attended in November 1998. Chief Gainer was officer-in-charge during the entire stand-off.

Very briefly, this dilemma centers around the length of the Roby stand-off. It was certainly the longest stand-off in the state's history, clocking in at 39 days. And it was a stand-off involving only one person. That person was a Mrs. Shirley Allen, a widow who had lived the life of a loner since the death of her husband about 10 years earlier. She was a person who behaved erratically and was convinced that aliens were out to get her. After attempting to visit her in the fall of 1997, and after being threatened with a shotgun, members of her family asked the Illinois court to involuntarily commit Mrs. Allen for her own protection. The court agreed that there was sufficient evidence and signed an order that she be detained. Three county deputies (together with Mrs. Allen's neighbor and a brother) went to the home to pick her up for an appointment with the mental health authorities. Mrs. Allen responded by coming out of the house pointing a 12-gauge shotgun. The deputies called for help and the Illinois State Police tactical team responded. To cut a very long story short, no fewer than 39 days later the stand-off came to a peaceful end when Mrs. Allen was apprehended, but not before the tactical team had experienced some very hairy moments!

Chief Gainer's talk on the episode was an enlightening and revealing one. He talked of his loneliness as officer-in-charge and of the criticism and "heat" he and his department took constantly from the media and other places for his department's unwillingness to "blow this old woman away," and for the enormous costs that built up during the 39 days that had to be borne out of the force's budget.

Questions

1. Can the subsequent dollar cost of this exercise be justified by the organization resolving the situation peacefully and without loss of life?
2. Is it possible to put a dollar value on a human life, no matter how mixed up or "crazy" that life may be?
3. Had you been the commander of this operation, keeping in mind the enormous media and political pressure to bring this situation to an end, what limits do you think there would have been to your patience?
4. Do you think the commander's handling of the situation would have been different if Mrs. Allen had been a man?
5. Would your thinking be different if the stand-off had involved a man?
6. I am going to stick my neck out here. Feel free to have a go at it if you feel the need! After the chief's presentation, I said the following: "On behalf of the general population, and I speak as one of only a few civilians in this room, let me thank you, as a civilian, for the value you placed on a human life in that situation." What do you think of my response?
7. Would it matter if the person was crazy? Does this somehow lessen the value of a human life?

Footnote: As I write this dilemma in the spring of 1999, an interesting inquest is under way in Toronto. It involves the fatal shooting by police in 1977 of a schizophrenic street person on a parked bus in Toronto. His name was Edmond Yu. Three police officers were present on the bus, one of them off duty, and the unarmed man was shot six times and killed. He had threatened the police with a hammer he had pulled out of his coat. One of the officers carried pepper spray, which was not used. The victim was a former medical student at the University of Toronto who had succumbed to his schizophrenic demons and had started to lead a bizarre sort of street life. The police have and are enduring relentless criticism from the public and the media for what many are perceiving to be an overly hasty, and lethal, response to the problem.

Dilemma No. 14

A Terrible Compulsion

You are a corrections officer in a minimum security facility for young offenders. You are working the night shift when a youth is admitted. This particular resident is very tall, thin, almost emaciated, and very frightened to be in custody. You begin to feel some empathy for this very awkward kid because this is the first time he has been in custody and is clearly scared out of his wits. He is placed in secure isolation for the rest of the night because he told the staff he is feeling suicidal.

A few days later he is integrated into the general population and pays a great deal of attention to you, presumably because he sensed your empathy on the night he was admitted. Over the course of a few days, news about the kid begins to emerge. You look through his psychiatric file to check for yourself what your colleagues are talking about and sure enough the file confirms that this youth does not wish to remain a male, rather he wants to be female.

One day while doing a routine cell block search you enter this youth's cell to find him stuffing something into his pillowcase. You tell him you saw him doing this and he looks terribly embarrassed, scared, and panicky. You ask him to show you what is in his pillowcase; he reaches in and brings out some women's panties. With a pleading look he explains that he prefers them to boxer shorts. He begs you not to mention this to other staff, that it would save him much hassle and mocking. You tell him to finish cleaning his cell. It is now time for lunch and you have to go and pick up one of the meal carts.

On the way down to the kitchen the scenario played over and over again in your mind. No other officer witnessed the event. This kid is obviously very mixed up, and telling the other officers might make him more of an outcast than he already is. You think to yourself, "What is the big deal about underwear anyway? But, on the other hand, if it ever came to light that I knew about his tendency and did not report it to a senior officer, my credibility would be shot and I would look like a fool."

Questions

1. How much weight, if any, would you place in your decision making about whether or not to say anything about this incident, on the youth's embarrassment and plea for privacy?

2. Assume for a moment that you have one or two colleagues who would make fun of this kid because of his predilection for wearing women's underwear. Should this fact play a part in your decision making?
3. Which of the tools, if any, may help you with this dilemma?

Dilemma No. 15

Different Standards

You are residential counselor at a wilderness facility for hard-to-manage teenage offenders. You have been in your job for three months and you are very much enjoying the experience.

It has taken a while, but after you have been at the camp for a couple of months you are invited to a social evening (party!) at the home of one of the other counselors. These staff get-togethers are held fairly often. After all, this is a wilderness camp set miles from anywhere and even the staff homes are very isolated. The nearest town of any size is many miles away.

You have gotten to like your colleagues very much. They are highly skilled in working with the youth, they are generally warm and friendly, particularly to you, and they perform their work with the kids as if it is a vocation. So you are especially pleased to feel you have been included in the group and have been invited to one of their parties.

The night of the party has arrived and you drive to the home where it is going to be held on this particular occasion. The camaraderie was great and this, you thought, was a terrific way for staff to relax and have fun together.

During the evening, though, the drugs came out. Not particularly hard ones, not particularly soft ones either. Everyone at the party proceeded to get stoned—except you.

What worries you more than anything, except perhaps the illegality of what was going on, was that these staff members regularly held formal discussions with the youth about the perils of doing drugs.

Questions

1. There are people who would argue that what went on in the privacy of this home should be of little or no concern to anyone else, including the police. What do you think?

2. There are people who would argue that these staff members are guilty of hypocrisy in that they preached one thing (perhaps only because they were paid to do so) and practiced another. How would you respond to this point of view?
3. There are those who argue that any person involved in a position of public trust (as these staff members were) should conduct their private lives accordingly, that is, conduct their private lives in a manner that is consistent with their public trust and responsibilities. What do you think of this idea?

Summary

These are just a few examples of the sorts of dilemmas and problems that can be encountered. My hope is that you will be challenged by them, but that at the same time you will enjoy the process of opening your mind to new and interesting angles and perspectives.

I would encourage you to generate your own scenarios, preferably drawn from your own experience, and to air them with fellow students.

In the meantime, here are some more brainteasers that you may want to ponder:

- Who should police the police?
- If a recruit officer is caught cheating on an examination during the last week of his training, should he be dismissed?
- If and when you are told by your coach officer that you should forget everything you were taught at the academy, college, or training center, what do you think you should do with this information?
- Is there a place for a whistle-blowing officer on any police force or in any correctional institution? If yes, under what circumstances? If no, why not?
- Are your colleague's attitudes, behaviors or lifestyle choices any of your business?
- What place is there for officer discretion? Is it possible for officers to use discretion, and still be considered fair?
- Dr. Martin Luther King said there are good laws and bad laws; good laws should be upheld, bad laws should be disobeyed. He was advocating the view that at times there can be a place for civil disobedience. How do you respond to this idea?

- Have you ever considered how you would respond if you were assigned a duty as a police officer that conflicted with your personal values? (For example, there was a case on a large metropolitan police force in which a veteran officer was assigned to ensure the safety of a doctor and his abortion clinic; the officer refused the detail on religious grounds, and was dismissed from the force.)
- One veteran police officer once told me that he knew of no police officer who had not, on at least one occasion, perjured himself. How do you respond to this point of view?
- Is there a place for complicity in report writing?
- Would you ever lie to save a fellow police or corrections officer?
- When you are being questioned about an incident to which you were a party, is the act of omitting important details that are known to you the same as lying? Or not?
- Should police forces and correctional institutions be open to public scrutiny? When asked by one researcher for access to his officers, one police chief responded, "The (name deleted) force has not enjoyed a particularly productive experience with sociological studies of various kinds." Should there be a place for independent, unbiased inquiry into the role of policing and corrections in our society?
- Should police officers who have been fired from their jobs for discreditable conduct be permitted to go into the private security industry?
- Should the testimony of jailhouse informants (who normally benefit personally from their willingness to give evidence) be permitted in court cases involving serious charges?
- If you were a police officer would you, at a friend's request, obtain confidential information for him from the police computer system on a former lover?

Reputable Officers

You see, what's happened is that now I've learned this new material [in the ethics course] I've become more accountable. It's pretty frightening when you pay $112 [course fee] and become more accountable. Generally you get paid for being accountable. There is obviously a certain responsibility that comes with an education—this one in particular.

—Journal entry, justice student in ethics course

Hopefully, as we go through life we will continue to grow and mature.

—Journal entry, justice student in ethics course

Perhaps this meditation [the ethics course] will make us realize that much of what we hold near and dear is not valid. What if I look deep within myself and find that I don't like what I see?

—Journal entry, justice student in ethics course

I believe that if I go to bed tonight knowing something that I didn't know this morning I am a better person.

—Journal entry, justice student in ethics course

To be part of the problem or part of the solution—that is the big question.

—Anonymous

The Point of the Exercise

I am going to assume that you have courageously had a go at some of the hard questions contained in this book; and that, either on your own or with the gentle or not-so-gentle prodding of your facilitator, you have been as honest as you can be. You may well now be asking questions: What has it all been for? What's the point? Has the exercise been worth it?

In the final analysis, of course, you must answer these questions for yourself, and come to your own conclusions. But in the meantime, let me suggest a few things to you.

First, perhaps it's better to think about the hard questions and tough issues raised in this book in the relative comfort and safety of a classroom, rather than when you are out on the street or on the institution range. I hope there will be some transference value to the thinking and discussing you have been doing during this learning process, so that the difficult decisions you will no doubt have to make on the job will be a little clearer, a little crisper, and, perhaps most important, ones with which you will feel comfortable. Even so, the feelings you will be left with after making these decisions, as I have suggested several times, may not be ones of complete peace, because the nature of a dilemma is often that there is no one perfect solution. More often than not, what we would be seeking to make would be the *best* decision rather than the *right* one. The decision you make may simply be the lesser of two evils, that is, that it will have a downside.

Second, if engaging in this exercise helps just one of you to make a decision that you may not otherwise have made, or helps to prevent you from making one that you may have made and regretted, then I would like to think my work in writing this book—and your effort and diligence in working through it—is justified. I'm thinking here of really important decisions—ones that will define you as a person, and will either contribute to justice, fairness, and decency, or detract from these values that we hold in such high esteem.

But whatever your view of these two points, don't relax just yet. We are not quite finished with this process! To conclude the book, there are a few additional points of discussion that I would like to engage you in.

At this point, I am going to express some personal, and perhaps controversial, thoughts on two questions. I express these views simply to encourage you to engage in further dialogue about them, and to leave you with some final ideas to think about. I claim no special wisdom in this regard, and there are no easy answers to these issues. But I am going to take a risk and be a little more forthcoming on these issues than I have been up to this point. I am going to lead with my chin, as it were, and set myself up on a pedestal. Of course, you are quite free to knock me off my perch if you can—so long as you can defend your views. Try to do so, but remember to consider the tools we discussed in Chapter 8. Check to see if they can support your arguments.

Can an Independent Thinker Be a Team Player?

One question that you may have been asking yourself—or perhaps the topic has come up in class—is "If I seek to be my own person as a police officer or corrections officer, how is this going to affect my ability to operate as a team player? Isn't being able to be a team player critical to the role of a police or corrections officer?"

This is a good question, and it is one that needs to be asked. You are correct if you think that this book essentially is about the individual officer, and the ability of that individual to make independent moral decisions and to have the courage to act on those decisions. You are also correct if you think that there may be some implications, in holding this view, for the officer group.

Let me suggest that there is something deeply troubling, and potentially dangerous, about individuals who are prepared to give up their sense of self and their individuality simply to ensure they fit into a particular group. Let me further suggest that if any group expects of any one of us that sort of mindless, unthinking obedience, then surely we should seriously question whether we wish to continue our association with that group.

Now of course I am not suggesting that a corrections officer or police officer group is normally characterized by such an expectation. I am suggesting, however, that there may be moments when we feel that that expectation is very much there. And it is at those moments, as rare as they may be, that how we respond is critically important to our welfare and well-being.

So what are we to do with those odd moments when we feel group pressure bearing down on us to participate in an activity with which we feel considerable discomfort? Is it possible at those moments to be our own person and still continue to be accepted and respected as a team player?

I think it is. And I think so because I have been privileged to meet several officers who have been able to keep that important distinction in their professional lives. The ideas I am going to suggest here, I have learned from these officers.

First, I think we owe it to each other as members of a colleague group to be up front with each other. When I ask people in my seminars what names they have for officers who break ranks and report a colleague, the word "snitch" is one of the first ones mentioned. "Snitching" is a pejorative word because it implies our running behind a person's back to tell something to someone, normally a person in authority. It implies acting in an underhanded way. Often this can be destructive, demoralizing behavior, and I would question its value or propriety in the context of policing or corrections. Something similar can be said, of course, of other professions. But in the context of policing and corrections, where being there for each other is so important, and where trust is critical, it is imperative that we seek ways to let our colleagues know exactly how we are responding personally when something is bothering us. This will require moral courage. It will be risky. But would this approach not be a better option than choosing to be devious and underhanded in the way we address what we consider to be a wrongdoing?

Of course, there may be times when we may be so troubled by something that is happening, and so terribly scared of the possibility of colleague retribution, that the only solution available to us is to go and speak with a senior officer in confidence. That is an option that each of us can consider when we feel we are caught.

How the senior officer responds to your request for anonymity will be interesting for you to discover. If he or she guarantees you anonymity and then you deliver some information about which she simply must do something, this will present a rather obvious dilemma for that officer. There's the commitment made to you by her versus her professional and ethical responsibility to do something with the new piece of information. If the manager feels she cannot guarantee you anonymity, and she tells you this at the outset, then, of course, she runs the risk of not finding out something that is clearly important.

What would you do in this situation if you were the manager? Would you commit to holding the information and the person conveying it in complete confidence, prior to knowing what the information is? Let's continue our discussion.

What we need to aim for, I would suggest, is to establish a climate in a policing or corrections context where we, as officers, give each other permission to approach each other in an open and honest way when something is not sitting well. If you feel this is a kind of pipe dream, something that is completely unrealistic, then we need to ask ourselves the question "Why would it be so unrealistic to expect this kind of agreement between a group of people who have sworn to uphold law and order?" What about those situations, you may ask, where we feel that simply discussing the situation with our colleague is not going to be enough. The situation is serious enough that we feel the truth must come out. Well, here I can say to my colleague something like this: "I feel really bad about your putting your uniform on when you're not on shift to get free meals from the Chinese restaurant, and I think it is wrong of you to do that; we're not supposed to take gratuities when we are on shift, let alone do what you are doing. It's up to you, I have spoken with you several times about this, and you have not taken me seriously. Either you come clean about it, or you will leave me no alternative but to report it. You are making all of us look bad. I can't stay quiet about this."

The officer in question, you will have noted, is not only compromising himself and his force by accepting gratuities, but he is also gaining the gratuities by deceitful means. There are two wrongs here.

I told you I was going to lead with my chin, didn't I? Would it be a fair assumption that this suggestion is eliciting a fairly strong emotional response in you? *No?* That's interesting. *Yes?* Why do you think you are reacting as you are?

If you are reacting negatively to what I have suggested, perhaps you can tell me how else we should approach issues of wrongdoing? If I remain silent about my colleague's inappropriate conduct I am, in effect, condoning it. Some folks have suggested that silence implies consent, and that if I know of a wrongdoing and keep quiet about it I am equally guilty.

What obviously gets in the way of justice being served at these moments is my feelings for, and my loyalty to my colleagues, right? These are noble sentiments, and such loyalty and concern for each other in our dangerous workplace is so important. But what if it leads to abuse, corruption, cover-up, injustice, and betrayal of public trust? What then? What about the public you serve and the trust they have placed in you as a peace officer to do the right thing? What about your own commitment to honesty, impartiality, fairness, justice, equity, and human rights?

Another important question is this one: Is there a possibility that if educators introduce you, the student, to this ethics experience, and it serves to reinforce your moral sense and contributes to your becoming keener than ever to fulfil your duties in a highly principled way, it might set you up for disappointment and bewilderment once you join the ranks of the officer group?

As the first edition of this book was being written in mid-1997, there was an inquiry into police corruption in New South Wales, Australia. As a result of what can only be described as a scathing indictment of the force by a royal commission, new police recruits in that state are now required in their training to participate in an ethics educational experience.

In May 1997, the first 219 police graduates of the new training program were sworn in as probationary constables. "Class 268 had learnt about ethics and the 'highest possible standards' during their time at the academy, and they 'must never tolerate, never ignore' corruption, the abuse of police powers or poor service delivery," said Peter Ryan, the police commissioner. One graduate, in an interview is reported to have said, "Just do the right thing and keep my nose clean—and whatever you see, you have to report it." Another graduate said, "I think those [corrupting] temptations are always there, but what they're trying within the recruiting system now is basically to be able to avoid that temptation, and look at moral strengths among their recruits rather than selecting people that may be swayed" (*The Sydney Morning Herald*, May 17, 1997). Is this idealism, as laudable as it is, inherently risky or even dangerous for the new recruit? The group, as we have suggested, can be set in its ways, and is often characterized by the norm that the officer's commitment to his colleague group should be honored above all else.

This question about whether we may be setting up recruits for failure is a good one, and there is no easy answer. We clearly do not wish you, as new officers, to leave your educational experience like white knights riding all-knowing chargers, acting as though you have a lock on morality. The officer group has its own ways of dealing with new officers who come on in this way, and the correcting experience can be a painful one!

Let me suggest, however, that any ethics educational experience within the context of a justice program would be incomplete without talking about the officer group and its dynamics, including the subcultural constraints issue. I remember one graduate of a two-year police education program telling me that there had been no mention at all during her studies of what she might have anticipated in terms of colleague group pressures.

Perhaps it is only by fully informing you about the subculture and how it may affect you—the new officer or would-be officer—that you will be prepared for what you may expect on the job. Then what you do with your moral idealism is up to you. Somehow you will have to look for ways to quietly and humbly convey your personal standards in a way that is not going to get you on the wrong side of your colleagues.

Look around you, see who you admire. Who best exemplifies a sense of true professionalism and decency in the way he does the job? Model yourself on that officer. Tell him of your respect, and seek his advice and help.

Should We Admit to Wrongdoing?

What do you think is the best policy: to admit to a wrongdoing one has committed, or not? I guess for almost every person who chooses not to admit to a wrongdoing and is subsequently found out, there are an equal number of people who benefit from covering up their wrongdoing. I have no statistics to support this view, and I may be wrong, but that's my sense of the way things are.

If I am correct in my assumption, then there would appear, at first blush, to be some advantage in at least attempting to cover up a wrongdoing, right? This would apply especially in cases where it is known that it is going to be difficult for the investigators to produce evidence of the wrongdoing.

Many of us seem to have a natural tendency to deny responsibility when we are confronted about a wrongdoing; perhaps this kind of defensiveness has its roots in our instinct for survival. There are probably few of us who have not engaged in lying in an attempt to save our skins. I am as guilty as anyone else in this regard. Would you agree that these moments when we engage in this kind of defensiveness and deceitfulness are moments that are not easily forgotten?

There are moments in each of our lives that I call defining moments. They are made up of experiences that deeply affect us in one way or another, for good or ill, and that leave us changed in some way.

Let me be quite personal with you at this point as I tell you about one such defining moment in my life. I grew up in rural Wales. I lived in a tiny village called Little Honeyborough. The village consisted of 10 cottages, two farms, and a chapel. As a child I was encouraged (no, told!) by my parents to go to the chapel every Sunday. There wasn't much of a protest from my brothers and sisters and me, because when I think about it now, the chapel was the social center of the village. (I can imagine you muttering as you read this, *Get a life!*)

One big event of the year was the Sunday School fete. It was a fundraising, fun event, with various market stalls and innocent games booths. One year—I think I must have been about 10 years of age at the time—it was my job, along with another Sunday School "scholar," to operate a booth where the idea was to throw table-tennis balls into goldfish bowls, much like you would see at some fairs now. For threepence (a few cents) a go you could have three shots at winning a goldfish!

During the day, my fellow helper excused himself for a bit, and I was left on my own to operate the stall. It was at this point that the defining moment took place. I helped myself to (stole) half a crown (two shillings and sixpence), worth about 50 or 60 cents now, but worth much more then. The fact that the cash tin was short was noticed by my colleague when he returned. He spoke with me first, and then to one of the organizers, about the missing money. I was questioned regarding the missing money, and of course I denied knowing anything about it.

The guilty feelings associated with that incident remain with me to this day, many years later. This was a really defining moment in my life even at that early age, because it forced me to face up to the struggle between right and wrong. It also introduced me, perhaps in a way that hadn't happened to the same extent prior to that experience, to the anguish of deep guilt. In retrospect, I regretted then, and regret now, not admitting to taking the money. I often wonder how this event may have affected me differently had I done so. In any event, there is no doubt in my mind now that the nonverbal signals I would have been giving off during the discussions about the missing money would have literally screamed that I was the guilty party. So not only would the organizers have known that I had taken the money, but they would have known I was a liar.

It's interesting, but even though I think I learned much from what may appear to others to be a rather innocuous experience, the temptation is still very much there to lie or to be evasive when confronted. It is almost as though the defensive reaction is a primeval one. My wife would no doubt confirm this for you! Am I alone in exaggerating traffic problems when late for dinner? Or in coming up with some ingenious "reason" (and it would have to be ingenious) why an important occasion has been overlooked?

Now let's attempt to relate these meandering thoughts to life as a police or corrections officer.

If I make a mistake—if I screw up, to use the vernacular—should I admit to my wrongdoing? In the final analysis, of course, we have to make our own decisions. No one should or can make them for us.

I pull the trigger too hastily in my panic; I fall asleep on night shift, and due to my neglect an inmate takes the opportunity to hang himself; I allow myself to become tunnel-visioned in a case I am investigating, and because of my investment in the case I refuse to consider new evidence that indicates the accused is the wrong person; I allow myself to be duped into taking a "compassionate" letter out of the institution for an inmate, the contents of which subsequently contribute to drugs being brought into the facility.

During the investigations that will inevitably take place into incidents like these, should I be honest and aboveboard, or should I attempt to get myself out of the tough spot by lying?

I heard an interesting discussion about whether or not we should own up when we have made a mistake. A senior police officer described something that had happened on his force. One of his officers shot and killed a young male suspect. The officer was accompanied by two colleagues. The case went to trial and the officer was acquitted of the shooting. A few months later he retired. Several weeks after that he came forward to admit that he had lied, as had his two colleagues. Apparently the officer could not live with his guilt, which is why he turned himself in.

Then the senior officer who was recounting the story said: "I think he is going to be okay; at least he is going to be better off than the two colleagues who lied and did not own up to their lies. They will certainly be going to jail."

Whether or not to own up to wrongdoing is a difficult question, and one you must address to yourself. Do some thinking about it now, so that you may be better prepared for struggling with it if and when the time comes.

Owning Responsibility

Meanwhile, let me offer you some thoughts on this question about owning responsibility for wrongdoing. Again, these ideas are offered simply in the interests of sparking some of your own thinking and discussion.

Would you agree that we are, generally, much more ready to forgive someone who has made a mistake, and who admits to that mistake, than we are to forgive someone whom we think is clearly guilty and whom we feel is lying and cheating? This is a rather curious response, if I am right in thinking that most of us instinctively, when we have done something wrong, feel a great temptation to get into a defensive mode. Maybe this is the reason we think it is so laudable when some individuals can overcome the natural urge to lie, and be big enough to accept responsibility for their actions.

As human beings, we are all too familiar with our predilection for making mistakes, and with our tendency from time to time to engage in wrongdoing. But even when the wrongdoing is serious, if the person making the mistake is courageous enough to admit to it, we can be forgiving. This is probably because we identify with the person as one imperfect human being to another.

You may remember one of the dilemmas we discussed (Number 4), in which an officer had uncharacteristically hit a young offender and was subsequently contrite about his indiscretion. Didn't you feel for the officer when you were discussing the dilemma? And what about the two students who were brought to me for cheating on an examination (Number 1)? One student, you may remember, was very defensive about the whole affair, and the other was ashamed and embarrassed about what he did. Doesn't a person's subsequent response to his mistake make a difference to the way we respond to him?

I think it would be wise to offer a word of caution here, however. A false act of "contrition," an apology or an accepting of the responsibility for a wrongdoing that is offered without sincerity, is, without putting too fine a point on it, manipulative. As humans, we generally have our antennae out for this kind of manipulative behavior, and it often makes us much less forgiving of the wrongdoer.

One of my students recounted an event she remembered when she was about seven years old. It is significant, I think, that this young woman still had a clear recollection of this event some 15 years later. What is also interesting to note is that she felt much guilt for her wrongdoing, and much relief subsequently, for owning up to it.

"My sister and I were sitting at the table eating lunch and I didn't want to drink my milk," she said. "Instead of telling my mom I didn't want it I tipped it over and spilled it and told my mom that my sister had done it on purpose. I guess I was jealous and wanted the attention, I don't remember. I do remember the way I felt when my mom started yelling at my sister and threw her into the big brown chair. I had the sickest feeling in my stomach and I felt so bad about what I had done that I went over to the 'quiet chair' with my sister (I guess to punish myself) and cried right along with her. I then called my mom over to tell the truth. I felt so relieved after doing the right thing." And her mom forgave her— presumably for owning up to her wrongdoing and telling the truth. Would you agree that we seem to have a sixth sense about whether someone is truly contrite or not? Of course, there will always be occasions when we can be duped, but normally we seem to know instinctively if a person's response is genuine.

The law enforcement field is replete with examples of the good that comes from owning up to one's mistakes. Several years ago, for example, the Boston Police Department had a drug warrant for a suspect, and information from an informant that the culprit was present in a third-floor apartment in a housing project. The police department formulated a plan of attack, brought in the SWAT team, and at the appointed time, took down the door of the apartment and went charging in. As they entered, they saw a man seated in the living room jump up and run into the bedroom. Two officers ran after him, apprehended him, took him to the floor, and handcuffed him. The officers immediately noticed that the man was vomiting, at which point they released the restraints and called for medical assistance. Within three minutes EMTs were helping the man, but it was not enough, and the fellow died. As it turned out, the man who died was a 74-year-old retired minister—the police had entered the wrong apartment, and the elderly gentleman died from the shock of the entry.

In response to this tragedy, the Boston police took a fairly uncustomary route. They stepped forward and publicly announced: "We were wrong! We made a mistake! We did not intend that the Reverend die, but it was a mistake of the mind and not of the heart." The police commissioner attended the Reverend's wake, he went to the funeral, and he oversaw a very public investigation of the event to find ways to, as best as possible, make certain such a terrible event would not repeat itself.

When the Boston police chose to accept responsibility for their mistake, there was a very interesting response from the community. While clearly unhappy that an innocent man had died as a result of police action, the forthright admission of error (coupled with overt steps to prevent similar problems in the future) tended to ameliorate much of the outrage that often follows such events.

A few years ago there was a heavyweight boxing match between Mike Tyson and Evander Holyfield. Some of you may remember that during the fight Tyson bit both of Holyfield's ears, tearing a small piece out of one of them. The fight was stopped by the referee, who disqualified Tyson. At a press conference, Tyson read what appeared to be a carefully scripted statement saying he was sorry for what he had done. But in my view, there was no *heart* in what he said. There was little of him as a human being in his words—probably because the words he read were not his own. It should not be surprising that many folk have since raised an eyebrow over this act of "contrition," and remain unconvinced that Tyson really felt that he did anything wrong.

While we are discussing sport (am I displaying my bias here?), at the Pan American games held in Winnipeg, Canada, in August 1999 a Canadian athlete was disqualified for using no fewer than three banned substances. The Canadian roller hockey team had just won the gold medal. As a result of the athlete's indiscretion the entire team and its coaches were stripped of their gold medals. As you can imagine this player's colleagues were not too happy with him. Compounding their unhappiness, however, was the "apology" this individual offered at a press conference called a few days after his disqualification and after he came out of hiding. The apology was a grudging one, an apology but not an apology. A *Toronto Star* article (August 4, 1999, p. E7) carried the headline "Vezina Teammates Angrier After 'Apology'." Said one of his teammates, "I think right now I am more angry than I was before. . . . He basically claims stupidity. . . . He's claiming he didn't know what drugs were illegal. That's a crock. We knew. We had our own orientation video. As far as I'm concerned he's an absolute liar. He's not just lying to himself, he's lying to the whole country and that's a shame."

So would it be fair to say that as humans we do not respond well to people who attempt to lie their way through an investigation of wrongdoing? And that we do not respond well to people who, having been found to have committed a wrongdoing, indulge in what appears to be a fake act of contrition?

A few years back, a female first lieutenant in the U.S. military faced a court martial on a charge of having committed adultery while serving as an officer in the military. Lieut. Kelly Flinn, America's first woman B-52 bomber pilot, received what is called a "general discharge" from the military. This book is not the place to debate the U.S. military's policies around officers' sexual conduct, but it is interesting to note that senior military personnel cited Lieut. Flinn's alleged initial dishonesty and denial as one of the mitigating factors in their decision to discharge her from the service. That would seem to imply that had there been a more straightforward, honest initial response from her, the decision might have been a different one.

A Human Response to Human Weakness

I can't tell you what you should do at those moments when you find yourself on the hot seat, just as you can't tell me. But perhaps it would be worth keeping in mind how positively we generally respond to people who accept responsibility for their mistakes, and how negatively we generally respond to people who are evasive.

There is one additional factor to think about. We should consider how we would feel about ourselves if we managed to get away with a wrongdoing by lying about it. If we are able to take comfort from, or gloat about, such a hollow victory, perhaps we should think seriously about whether or not we continue to be worthy of the important law-keeping role in society we enjoy or are seeking.

One officer I once spoke with, who had been reprimanded for committing a serious indiscretion about which he had responded honestly, said, "At least I can live with myself; I knew what I did was wrong, but I felt that I had to take responsibility for it." When this person said this, I have to say he went up in my estimation for the mature way in which he accepted ownership of his actions. Although the act committed clearly would meet with our disapproval, would you not agree that this officer's response is one to be admired?

Let's draw to a conclusion these thoughts on whether or not it is wise to admit to wrongdoing. Today as I write these notes, the report of the inquiry commission investigating accusations of wrongdoing by the Canadian military in Somalia (referred to in Chapter 1) was published. Here are some comments made by the commissioners about two of the principal players in that affair. Of Gen. Jean Boyle, the commissioners report: "We are satisfied . . . that Gen. Boyle was a party to the decision to informally release altered documents . . . and that he knew of such alterations. We are satisfied . . . that Gen. Boyle participated in the devising of a process which provided the public with misleading or incomplete information and condoned such a process." But of Maj.-Gen. Lewis Mackenzie, they had this to say: "Gen. Mackenzie is unique among the leaders who appeared before us. . . . [He] testified before us in an honest and straightforward manner. He alone seemed to understand the necessity to acknowledge error and account for his personal shortcomings. . . . His comportment and demeanour throughout his testimony before us were consistent with the highest standards of military duty and responsibility" (*The Globe and Mail,* July 3, 1997, p. A4).

What is important to note is that the commissioners did not find Maj.-Gen. Mackenzie blameless. He had made serious mistakes. The commission found that "Maj.-Gen. Mackenzie's fundamental failing was that he exercised inappropriate control and provided inadequate supervision" (*The Globe and Mail,* July 3, 1997, p. A4). He made his mistakes, but what separated him from the other leaders in the minds of the commissioners was his willingness to accept responsibility for those mistakes.

In the United States, there is a simple word that accurately describes public attitudes about the continued travails of President William Clinton. The word is fatigue. Many people have reached the point where continued discussion and examination of his interactions (or lack of same) with volunteer White House staff have become, at the very least, wearisome. Having long since learned not to expect a straight answer to the "did he or didn't he?" question, most citizens have also lost any hope for a sincere expression of personal accountability. Instead of accepting responsibility for his actions, Bill Clinton treated us to a rare glimpse into the netherworld of legal maneuvers, with the capstone certainly being his sworn deposition in which he declared: "It depends on what the meaning of the word is is." How different and refreshing would things have been over the past couple of years had Bill Clinton chosen, early on, to stand up before the television cameras and say "I'd like your attention for a few moments. I have an announcement to make. I am a flawed individual, and I have done some things for which I am very ashamed, and I will take whatever steps are necessary to be sure such things do not happen again."

I offer these examples as support for the personal opinion I have been expressing: we can be forgiving of the wrongdoings and mistakes of others. This is especially the case if we feel that they are willing and prepared to respond honestly, and to take personal responsibility for their mistakes.

Please do with this discussion what you will. Think about it. Think about how you respond to people you have known who have been big enough to admit a mistake or a wrongdoing. Think about those individuals you have judged to be guilty who continue to deny responsibility. Think about manipulative "contrition."

I would encourage you to incorporate that thinking into your own personal decision making as you do the important work you do.

A Matter of Public Trust

If you are reading these final paragraphs, you have done some of the hard thinking demanded of you by this text. You are to be commended for accepting the challenge.

Perhaps your experience, like that of many of my students, has been that the process of seeking to answer the questions posed left you with a headache; that there were classes and group discussions that you left with your head spinning. Good! Your brain is a muscle just like your other muscles, and if your head ached it probably means that your brain had a bit of a workout. Or perhaps your headaches were caused by the tension you experienced because your values were challenged. I am pleased you were courageous enough to complete the exercise.

It is my sincere wish that you have benefited from this process in some small way, and that it has contributed to your development as a human being. It is also my sincere wish that you will do well in your career as an officer—the public you serve is depending on you. You will enjoy—as a relatively new officer or when you are hired—an enormous amount of public trust. Keep that trust sacred, and always seek to make the best decisions you can.

You will have noticed that I used the word *best.* I could have used the word *right.* If you know what the right thing to do is, and you do it, you are to be congratulated. But there will be other issues and dilemmas with which you will be confronted, which will not be quite so clear-cut. Here you will have to do what you can to make the best decision under the circumstances. At these moments those exercises may be of some help to you.

I wish you well. Be a good person. Be a good officer. Always do what you consider to be the right thing, or the best thing under the circumstances, as the case may be. Uphold the law, and what you consider to be the truth; seek to carefully and honestly consider where the truth lies. Be consistent. Be fair. Beware of colleagues who may seem to have lost their moral sense, and who aim to drag you down with them. Apply the law fairly and impartially without fear or favor.

Above all, in whatever you do and wherever you go, seek at all times to be your own person. Work out for yourself the place that colleague loyalty shall have in the way you do your job, and what the limits of that loyalty are for you.

We have two black-and-tan basset hounds at home. We adopted them in a moment of weakness and have loved them ever since! Suzie is the brains of the outfit, and Sam is as thick as two short planks. But both are lovable. One of the selling features brought to our attention by the previous owner was that Sam "sings." One mention of that word and he develops a funny look on his face, contracts his stomach and, 10 or 20 seconds later, emits a sound that would be enough to make even the most devout person doubt the existence of God. The interesting thing is that when Sam starts to "sing," Suzie starts to bark. The noise can be deafening. Our neighbors also have two dogs, and when ours bark, theirs bark. And if there were other dogs in the neighborhood, I have no doubt they would be at it, too. Why am I telling you all this?

Here's one last thought for you to take with you. Think about it, and work out for yourself what the message may be for us as human beings. There is an old Chinese proverb that goes something like this: One dog barks, and a hundred others bark at the sound.

As human beings, and as officers, are we that much different from our canine friends?

You be the judge.

References

Baker, Mark (1985). *Cops: Their Lives in Their Own Words.* Simon and Schuster Inc., New York.

Baron, Jonathan (1990). "Thinking About Consequences." *Journal of Moral Education,* Vol. 19, No. 2, 77–87.

Beck, Clive (1991a). "Approaches to Values Education." *Course Material: Learning Values in Adulthood,* Ontario Institute for Studies in Education, Toronto.

Beck, Clive (1991b). "How Adults Learn Values." *Course Material: Learning Values in Adulthood,* Ontario Institute for Studies in Education, Toronto.

Beck, Clive (1991c). "Moral Values." *Course Material: Learning Values in Adulthood,* Ontario Institute for Studies in Education, Toronto.

Bittner, E. (1980). *The Function of Police in Modern Society.* Oelegeschalger, Gunn, and Haire, Cambridge, Massachusetts.

Bok, Derek (1988). "Can Higher Education Foster Higher Morals?" *Business and Society Review,* Volume 66, 4–12.

Buscaglia, Leo (1978). *Love.* Fawcett-Crest, New York.

Callaghan, Daniel, and Sissela Bok, eds. (1980). *Ethics Teaching in Higher Education.* Hastings Centre on Ethics, Plenum Press, New York and London.

Cartwright, Darwin, and Alvin Zander, eds. (1968). *Group Dynamics: Research and Theory.* Harper and Row Publishers.

Chidley, Joe (1995, January 30). "Bonding and Brutality." *Maclean's,* p. 30.

Coffey, Alan, Edward Eldefonso, and Walter Hartinger (1982). *Human Relations: Law Enforcement in a Changing Community* (Third Edition). Prentice Hall, New Jersey.

Collins English Dictionary (Third Editon). 1991. Harper Collins Publishers, London.

Davis, Michael (1990). "Who Can Teach Workplace Ethics?" *Teaching Philosophy,* 21–38.

Desroches, Frederick J. (1986). "The Occupational Subculture of the Police." In: Brian K. Cryderman and Chris N. O'Toole, eds. *Police, Race and Ethnicity: A Guide for Law Enforcement Officers.* Butterworths, Toronto, 39–51.

Dimanno, Rosie (1995, January 30). Attacks on Kerr a Punishment for Breaking Ranks. *The Toronto Star,* p. A6.

Folse, Kimberly A. (1991). "Ethics and the Profession: Graduate Student Training." *Teaching Sociology,* Vol. 19, 344–350.

Fox, Vernon (1983). *Correctional Institutions.* Prentice Hall, New Jersey.

Goldstein, H. (1990). *Problem-Oriented Policing.* McGraw-Hill, New York.

Grossi, Elizabeth L., and Bruce L. Berg (1991). "Stress and Job Dissatisfaction Among Correctional Officers: An Unexpected Finding." *International Journal of Offender Therapy and Comparative Criminology,* Vol. 35, No. 1, 79.

Hess, Henry (1996, October 23). Former Constable Describes Tampering. *The Globe and Mail,* p. A9.

Horin, Adele. (1994, July 9). Why Loyalty is a Two-Way Street. *The Sydney Morning Herald.*

Janis, Irving L. (1968). "Group Dynamics Under Conditions of External Danger." In: Darwin Cartwright and Alvin Zander, eds. *Group Dynamics—Research and Theory.* Harper and Row, New York.

Josephson, Michael. "The Bell, the Book, the Candle." The Josephson Institute for Ethics, California.

Kauffman, Kelsey (1988). *Prison Officers and Their World.* Harvard University Press, Cambridge, Massachusetts.

Klofas, John, and Hans Toch (1982). "The Guard Subculture Myth." *Journal of Research in Crime and Delinquency,* July, 238–254.

Kohlberg, Lawrence (1984). *Essays on Moral Development, Vol. II: The Psychology of Moral Development.* Harper and Row, San Francisco.

Lamont, Leonie. (1997, May 17). Force's Hope for the Future Seeks a Keen Eye and a Clean Nose. *The Sydney Morning Herald,* p. 8.

Levy, Harold (1997, March 7). Courtroom 'Tense' for Detective in Shooting. *The Toronto Star,* p. A6.

Lickona, Thomas (1976). "What Does Moral Psychology Have to Say?" In: Lickona, Thomas, Gilbert Geis and Lawrence Kohlberg, eds. *Moral Development and Behaviour: Theory, Research and Social Issues.* Holt, Rinehart, Winston, New York.

Main, Eron. (1996, October 1). Discretion No Part of Valour. *The Globe and Mail*, p. A16.

Martinson, Barbara, ed. (1987). *A Discourse on Ethics and the Corporate Workplace: Can Ethics be Taught?* Working Paper No. 2. Corporate Council on the Liberal Arts, Working Paper Series, New York.

Mascoll, Philip (1996, November 21). Disciplinary Muscle Proposed for Chief. *The Toronto Star*, p. A6.

Millar, Cal, and Philip Mascoll (1995, March 17). Four Officers Suspended in Probe. *The Toronto Star*, p. A6.

Mills, Theodore (1967). *The Sociology of Small Groups.* Prentice Hall, New Jersey.

Noddings, Nel (1994). "Conversation as Moral Education." *Journal of Moral Education*, Vol. 23, No. 2, 107–118.

Paul, Richard (1992). *Critical Thinking* (Revised Second Edition). Foundation for Critical Thinking, Sonoma State University, Rohnert Park, California.

Sheehy, G. (1977). *Passages: Predictable Crises in Adult Life.* Bantam, New York.

Strike, Kenneth A., Emile J. Haller, and Jonas F. Soltis (1988). *The Ethics of School Administration.* Teachers College Press, New York.

Tafoya, William (1995). "Ethics and the Realities of Life: Surviving the Vortex." In: Daryl Close and Nicholas Meier, eds. *Morality in Criminal Justice.* Wadsworth Publishing Company, San Francisco.

Task Force on Ethics (1985). The University of Alberta. Alberta University, Edmonton. The Senate.

Task Force on Ethics (1988). Progress Review: The University of Alberta. Alberta University, Edmonton. The Senate.

Weber, Max (1967). In: Gerth, H. H. and C. Wright Mills, eds. *Essays in Sociology.* New York and Oxford University Press.

Index

A

B

C

D

E

F

G

H

I

J

K

L

M

N

O

P

R

S

T

U

V

W

Y

W